2005 Edition

ANSI/AF&PA NDS-2005

(revised standard)

Approval Date: January 6, 2005

Approved American National Standard

ANSI

N D S®

NATIONAL DESIGN SPECIFICATION®

FOR WOOD CONSTRUCTION

ASD/LRFD

National Design Specification (NDS) for Wood Construction 2005 Edition

First Printing: April 2005

ISBN 0-9625985-1-8

Copyright Permission
AF&PA American Wood Council
1111 Nineteenth St., NW, Suite 800
Washington, DC 20036
email: awcinfo@afandpa.org

Printed in the United States of America

FOREWORD

The *National Design Specification® (NDS®) for Wood Construction* was first issued by the National Lumber Manufacturers Association (now the American Forest & Paper Association) (AF&PA) in 1944, under the title *National Design Specification for Stress-Grade Lumber and Its Fastenings*. By 1971 the scope of the Specification had broadened to include additional wood products. In 1977 the title was changed to reflect the new nature of the Specification, and the content was rearranged to simplify its use. The 1991 edition was reorganized in an easier to use "equation format", and many sections were rewritten to provide greater clarity.

In 1992, AF&PA (formerly the National Forest Products Association) was accredited by the American National Standards Institute (ANSI). The Specification subsequently gained approval as an American National Standard designated ANSI/NFoPA NDS-1991 with an approval date of October 16, 1992. The current edition of the Standard is designated ANSI/AF&PA NDS-2005 with an approval date of January 06, 2005.

In developing the provisions of this Specification, the most reliable data available from laboratory tests and experience with structures in service have been carefully analyzed and evaluated for the purpose of providing, in convenient form, a national standard of practice.

Since the first edition of the National Design Specification in 1944, the Association's Technical Advisory Committee has continued to study and evaluate new data and developments in wood design. Subsequent editions of the Specification have included appropriate revisions to provide for use of such new information. This edition incorporated numerous changes considered by AF&PA's Wood Design Standards Committee. The contributions of the members of this Committee to improvement of the Specification as a national design standard for wood construction are especially recognized.

Acknowledgement is made to the Forest Products Laboratory, U.S. Department of Agriculture, for data and publications generously made available, and to the engineers, scientists, and other users who have suggested changes in the content of the Specification. The Association invites and welcomes comments, inquiries, suggestions, and new data relative to the provisions of this document.

It is intended that this Specification be used in conjunction with competent engineering design, accurate fabrication, and adequate supervision of construction. AF&PA does not assume any responsibility for errors or omissions in the document, nor for engineering designs, plans, or construction prepared from it. Particular attention is directed to Section 2.1.2, relating to the designer's responsibility to make adjustments for particular end uses of structures.

Those using this standard assume all liability arising from its use. The design of engineered structures is within the scope of expertise of licensed engineers, architects, or other licensed professionals for applications to a particular structure.

American Forest & Paper Association

TABLE OF CONTENTS

LIST OF TABLES

LIST OF TABLES IN THE NDS SUPPLEMENT

LIST OF FIGURES

1

GENERAL REQUIREMENTS FOR STRUCTURAL DESIGN

AMERICAN FOREST & PAPER ASSOCIATION

1.1 Scope

1.1.1 Practice Defined

1.1.1.1 This Specification defines the method to be followed in structural design with the following wood products:
- visually graded lumber
- mechanically graded lumber
- structural glued laminated timber
- timber piles
- timber poles
- prefabricated wood I-joists
- structural composite lumber
- wood structural panels

It also defines the practice to be followed in the design and fabrication of single and multiple fastener connections using the fasteners described herein.

1.1.1.2 Structural assemblies utilizing panel products shall be designed in accordance with principles of engineering mechanics (see References 32, 33, 34, and 53 for design provisions for commonly used panel products).

1.1.1.3 This Specification is not intended to preclude the use of materials, assemblies, structures or designs not meeting the criteria herein, where it is demonstrated by analysis based on recognized theory, full scale or prototype loading tests, studies of model analogues or extensive experience in use that the material, assembly, structure or design will perform satisfactorily in its intended end use.

1.1.2 Competent Supervision

The reference design values, design value adjustments, and structural design provisions in this Specification are for designs made and carried out under competent supervision.

1.2 General Requirements

1.2.1 Conformance with Standards

The quality of wood products and fasteners, and the design of load-supporting members and connections, shall conform to the standards specified herein.

1.2.2 Framing and Bracing

All members shall be so framed, anchored, tied, and braced that they have the required strength and rigidity. Adequate bracing and bridging to resist wind and other lateral forces shall be provided.

1.3 Standard as a Whole

The various Chapters, Sections, Subsections and Articles of this Specification are interdependent and, except as otherwise provided, the pertinent provisions of each Chapter, Section, Subsection, and Article shall apply to every other Chapter, Section, Subsection, and Article.

1.4 Design Procedures

This Specification provides requirements for the design of wood products specified herein by the following methods:

(a) Allowable Stress Design (ASD)
(b) Load and Resistance Factor Design (LRFD)

Designs shall be made according to the provisions for Allowable Stress Design (ASD) or Load and Resistance Factor Design (LRFD).

1.4.1 Loading Assumptions

Wood buildings or other wood structures, and their structural members, shall be designed and constructed to safely support all anticipated loads. This Specification is predicated on the principle that the loading assumed in the design represents actual conditions.

1.4.2 Governed by Codes

Minimum design loads shall be in accordance with the building code under which the structure is designed, or where applicable, other recognized minimum design load standards.

1.4.3 Loads Included

Design loads include any or all of the following loads or forces: dead, live, snow, wind, earthquake, erection, and other static and dynamic forces.

1.5 Specifications and Plans

1.5.1 Sizes

The plans or specifications, or both, shall indicate whether wood products sizes are stated in terms of standard nominal, standard net or special sizes, as specified for the respective wood products in Chapters 4, 5, 6, 7, 8, and 9.

1.6 Notation

1.4.4 Load Combinations

Combinations of design loads and forces, and load combination factors, shall be in accordance with the building code under which the structure is designed, or where applicable, other recognized minimum design load standards (see Reference 5 for additional information). The governing building code shall be permitted to be consulted for load combination factors. Load combinations and associated time effect factors, λ, for use in LRFD are provided in Appendix N.

Except where otherwise noted, the symbols used in this Specification have the following meanings:

A = area of cross section, in.^{2}

A_m = gross cross-sectional area of main wood member(s), in.^{2}

A_n = cross-sectional area of notched member, in.^{2}

A_s = sum of gross cross-sectional areas of side member(s), in.^{2}

C_D = load duration factor

C_F = size factor for sawn lumber

C_L = beam stability factor

C_M = wet service factor

C_P = column stability factor

C_T = buckling stiffness factor for dimension lumber

C_V = volume factor for structural glued laminated timber or structural composite lumber

C_b = bearing area factor

C_c = curvature factor for structural glued laminated timber

C_{cs} = critical section factor for round timber piles

C_d = penetration depth factor for connections

C_{di} = diaphragm factor for nailed connections

C_{eg} = end grain factor for connections

C_{fu} = flat use factor

C_g = group action factor for connections

C_i = incising factor for dimension lumber

C_r = repetitive member factor for dimension lumber, prefabricated wood I-joists, and structural composite lumber

C_{sp} = single pile factor for timber piles

C_{st} = metal side plate factor for 4" shear plate connections

C_t = temperature factor

C_{tn} = toe-nail factor for nailed connections

C_u = untreated factor for timber poles and piles

C_Δ = geometry factor for connections

COV_E = coefficient of variation for modulus of elasticity

D = diameter, in.

D_r = root diameter, in.

E, E' = reference and adjusted modulus of elasticity, psi

E_{min}, E_{min}' = reference and adjusted modulus of elasticity for beam stability and column stability calculations, psi

$(EI)_{min}, (EI)_{min}'$ = reference and adjusted EI for beam stability and column stability calculations, psi

E_m = modulus of elasticity of main member, psi

E_s = modulus of elasticity of side member, psi

F_b, F_b' = reference and adjusted bending design value, psi

F_{b1}' = adjusted edgewise bending design value, psi

F_{b2}' = adjusted flatwise bending design value, psi

F_{bE} = critical buckling design value for bending members, psi

F_c, F_c' = reference and adjusted compression design value parallel to grain, psi

F_c^* = reference compression design value parallel to grain multiplied by all applicable adjustment factors except C_p, psi

F_{cE} = critical buckling design value for compression members, psi

F_{cE1}, F_{cE2} = critical buckling design value for compression member in planes of lateral support, psi

$F_{c\perp}, F_{c\perp}'$ = reference and adjusted compression design value perpendicular to grain, psi

F_e = dowel bearing strength, psi

F_{em} = dowel bearing strength of main member, psi

F_{es} = dowel bearing strength of side member, psi

$F_{e\parallel}$ = dowel bearing strength parallel to grain, psi

$F_{e\perp}$ = dowel bearing strength perpendicular to grain, psi

$F_{e\theta}$ = dowel bearing strength at an angle to grain, psi

F_{rt}' = adjusted radial tension design value perpendicular to grain, psi

F_t, F_t' = reference and adjusted tension design value parallel to grain, psi

F_v, F_v' = reference and adjusted shear design value parallel to grain (horizontal shear), psi

F_{yb} = bending yield strength of fastener, psi

F_θ' = adjusted bearing design value at an angle to grain, psi

G = specific gravity

I = moment of inertia, in.^{4}

K = shear stiffness coefficient

K_D = diameter coefficient for dowel-type fastener connections with D < 0.25 in.

K_F = format conversion factor

K_M = moisture content coefficient for sawn lumber truss compression chords

K_T = truss compression chord coefficient for sawn lumber

K_{bE} = Euler buckling coefficient for beams

K_{cE} = Euler buckling coefficient for columns

K_e = buckling length coefficient for compression members

K_f = column stability coefficient for bolted and nailed built-up columns

K_r = radial stress coefficient

K_t = temperature coefficient

K_v = shear coefficient

K_x = spaced column fixity coefficient

K_θ = angle to grain coefficient for dowel-type fastener connections with D < 0.25 in.

L = span length of bending member, ft

L = distance between points of lateral support of compression member, ft

L_c = length from tip of pile to critical section, ft

M = maximum bending moment, in.-lbs

M_r, M_r' = reference and adjusted design moment, in.-lbs

N, N' = reference and adjusted lateral design value at an angle to grain for a single split ring connector unit or shear plate connector unit, lbs

P = total concentrated load or total axial load, lbs

P, P' = reference and adjusted lateral design value parallel to grain for a single split ring connector unit or shear plate connector unit, lbs

P_r = parallel to grain reference rivet capacity, lbs

P_w = parallel to grain reference wood capacity for timber rivets, lbs

Q = statical moment of an area about the neutral axis, in.^{3}

Q, Q' = reference and adjusted lateral design value perpendicular to grain for a single split ring connector unit or shear plate connector unit, lbs

Q_r = perpendicular to grain reference rivet capacity, lbs

Q_w = perpendicular to grain reference wood capacity for timber rivets, lbs

R = radius of curvature, in.

R_B = slenderness ratio of bending member

R_d = reduction term for dowel-type fastener connections

R_r, R_r' = reference and adjusted design reaction, lbs

S = section modulus, in.^{3}

T = temperature, °F

V = shear force, lbs

V_r, V_r' = reference and adjusted design shear, lbs

W, W' = reference and adjusted withdrawal design value for fastener, lbs per inch of penetration

Z, Z' = reference and adjusted lateral design value for a single fastener connection, lbs

$Z_\parallel$ = reference lateral design value for a single dowel-type fastener connection with all wood members loaded parallel to grain, lbs

$Z_{m\perp}$ = reference lateral design value for a single dowel-type fastener wood-to-wood connection with main member loaded perpendicular to grain and side member loaded parallel to grain, lbs

$Z_{s\perp}$ = reference lateral design value for a single dowel-type fastener wood-to-wood connection with main member loaded parallel to grain and side member loaded perpendicular to grain, lbs

$Z_\perp$ = reference lateral design value for a single dowel-type fastener wood-to-wood, wood-to-metal, or wood-to-concrete connection with wood member(s) loaded perpendicular to grain, lbs

a_p = minimum end distance load parallel to grain, in.

a_q = minimum end distance load perpendicular to grain, in.

b = breadth of rectangular bending member, in.

c = distance from neutral axis to extreme fiber, in.

d = depth of bending member, in.

d = least dimension of rectangular compression member, in.

d = pennyweight of nail or spike

d_e = effective depth of member at a connection, in.

d_n = depth of member remaining at a notch, in.

d_1, d_2 = cross-sectional dimensions of rectangular compression member in planes of lateral support, in.

e = eccentricity, in.

e_p = minimum edge distance unloaded edge, in.

e_q = minimum edge distance loaded edge, in.

f_b = actual bending stress, psi

f_{b1} = actual edgewise bending stress, psi

f_{b2} = actual flatwise bending stress, psi

f_c = actual compression stress parallel to grain, psi

f_c' = concrete compressive strength, psi

$f_{c\perp}$ = actual compression stress perpendicular to grain, psi

f_r = actual radial stress in curved bending member, psi

f_t = actual tension stress parallel to grain, psi

f_v = actual shear stress parallel to grain, psi

g = gauge of screw

ℓ = span length of bending member, in.

ℓ = distance between points of lateral support of compression member, in.

ℓ_b = bearing length, in.

ℓ_c = clear span, in.

ℓ_e = effective span length of bending member, in.

ℓ_e = effective length of compression member, in.

ℓ_{e1}, ℓ_{e2} = effective length of compression member in planes of lateral support, in.

ℓ_e/d = slenderness ratio of compression member

ℓ_m = length of dowel bearing in wood main member, in.

ℓ_n = length of notch, in.

ℓ_s = length of dowel bearing in wood side member, in.

ℓ_u = laterally unsupported span length of bending member, in.

ℓ_1, ℓ_2 = distances between points of lateral support of compression member in planes 1 and 2, in.

ℓ_3 = distance from center of spacer block to centroid of group of split ring or shear plate connectors in end block for a spaced column, in.

m.c. = moisture content based on oven-dry weight of wood, %

n = number of fasteners in a row

n_c = number of rivets per row

n_R = number of rivet rows

p = depth of fastener penetration into wood member, in.

r = radius of gyration, in.

s = center-to-center spacing between adjacent fasteners in a row, in.

s_p = spacing between rivets parallel to grain, in.

s_q = spacing between rivets perpendicular to grain, in.

t = thickness, in.

t = exposure time, hrs.

t_m = thickness of main member, in.

t_s = thickness of side member, in.

x = distance from beam support face to load, in.

α = angle between direction of load and direction of grain (longitudinal axis of member), degrees

β_{eff} = effective char rate (in./hr.) adjusted for exposure time, t

β_n = nominal char rate (in./hr.), linear char rate based on 1-hour exposure

γ = load/slip modulus for a connection, lbs/in.

λ = Time effect factor

ϕ = Resistance factor

2

DESIGN VALUES FOR STRUCTURAL MEMBERS

2.1 General

2.1.1 General Requirement

Each wood structural member or connection shall be of sufficient size and capacity to carry the applied loads without exceeding the adjusted design values specified herein.

2.1.1.1 For ASD, calculation of adjusted design values shall be determined using applicable ASD adjustment factors specified herein.

2.1.1.2 For LRFD, calculation of adjusted design values shall be determined using applicable LRFD adjustment factors specified herein.

2.1.2 Responsibility of Designer to Adjust for Conditions of Use

Adjusted design values for wood members and connections in particular end uses shall be appropriate for the conditions under which the wood is used, taking into account the differences in wood strength properties with different moisture contents, load durations, and types of treatment. Common end use conditions are addressed in this Specification. It shall be the final responsibility of the designer to relate design assumptions and reference design values, and to make design value adjustments appropriate to the end use.

2.2 Reference Design Values

Reference design values and design value adjustments for wood products in 1.1.1.1 are based on methods specified in each of the wood product chapters. Chapters 4 through 9 contain design provisions for sawn lumber, glued laminated timber, poles and piles, prefabricated wood I-joists, structural composite lumber, and wood structural panels, respectively. Chapters 10 through 13 contain design provisions for connections. Reference design values are for normal load duration under the moisture service conditions specified.

2.3 Adjustment of Reference Design Values

2.3.1 Applicability of Adjustment Factors

Reference design values shall be multiplied by all applicable adjustment factors to determine adjusted design values. The applicability of adjustment factors to sawn lumber, structural glued laminated timber, poles and piles, prefabricated wood I-joists, structural composite lumber, wood structural panels, and connection design values is defined in 4.3, 5.3, 6.3, 7.3, 8.3, 9.3, and 10.3, respectively.

2.3.2 Load Duration Factor, C_D (ASD only)

2.3.2.1 Wood has the property of carrying substantially greater maximum loads for short durations than for long durations of loading. Reference design values apply to normal load duration. Normal load duration represents a load that fully stresses a member to its allowable design value by the application of the full design load for a cumulative duration of approximately ten years. When the cumulative duration of the full maximum load does not exceed the specified time period, all reference design values except modulus of elasticity, E, modulus of elasticity for beam and column stability, E_{min}, and compression perpendicular to grain, $F_{c\perp}$, based on a deformation limit (see 4.2.6) shall be multiplied by the appropriate load duration factor, C_D, from Table 2.3.2 or Figure B1 (see Appendix B) to take into account the change in strength of wood with changes in load duration.

2.3.2.2 The load duration factor, C_D, for the shortest duration load in a combination of loads shall apply for that load combination. All applicable load combinations shall be evaluated to determine the critical load combination. Design of structural members and connections shall be based on the critical load combination (see Appendix B.2).

2.3.2.3 The load duration factors, C_D, in Table 2.3.2 and Appendix B are independent of load combination factors, and both shall be permitted to be used in design calculations (see 1.4.4 and Appendix B.4).

Table 2.3.2 Frequently Used Load Duration Factors, C_D[1]

Load Duration	C_D	Typical Design Loads
Permanent	0.9	Dead Load
Ten years	1.0	Occupancy Live Load
Two months	1.15	Snow Load
Seven days	1.25	Construction Load
Ten minutes	1.6	Wind/Earthquake Load
Impact[2]	2.0	Impact Load

1. Load duration factors shall not apply to reference modulus of elasticity, E, reference modulus of elasticity for beam and column stability, E_{min}, nor to reference compression perpendicular to grain design values, $F_{c\perp}$, based on a deformation limit.
2. Load duration factors greater than 1.6 shall not apply to structural members pressure-treated with water-borne preservatives (see Reference 30), or fire retardant chemicals. The impact load duration factor shall not apply to connections.

2.3.3 Temperature Factor, C_t

Reference design values shall be multiplied by the temperature factors, C_t, in Table 2.3.3 for structural members that will experience sustained exposure to elevated temperatures up to 150°F (see Appendix C).

2.3.4 Fire Retardant Treatment

The effects of fire retardant chemical treatment on strength shall be accounted for in the design. Adjusted design values, including adjusted connection design values, for lumber and structural glued laminated timber pressure-treated with fire retardant chemicals shall be obtained from the company providing the treatment and redrying service. Load duration factors greater than 1.6 shall not apply to structural members pressure-treated with fire retardant chemicals (see Table 2.3.2).

2.3.5 Format Conversion Factor, K_F (LRFD only)

For LRFD, reference design values shall be multiplied by the format conversion factor, K_F, specified in Appendix N.3.1. The format conversion factor, K_F, shall not apply for designs in accordance with ASD methods specified herein.

2.3.6 Resistance Factor, ϕ (LRFD only)

For LRFD, reference design values shall be multiplied by the resistance factor, ϕ, specified in Appendix N.3.2. The resistance factor, ϕ, shall not apply for designs in accordance with ASD methods specified herein.

2.3.7 Time Effect Factor, λ (LRFD only)

For LRFD, reference design values shall be multiplied by the time effect factor, λ, specified in Appendix N.3.3. The time effect factor, λ, shall not apply for designs in accordance with ASD methods specified herein.

2

DESIGN VALUES FOR STRUCTURAL MEMBERS

Table 2.3.3 Temperature Factor, C_t

Reference Design Values	In-Service Moisture Conditions[1]	C_t		
		T≤100°F	100°F<T≤125°F	125°F<T≤150°F
F_t, E, E_{min}	Wet or Dry	1.0	0.9	0.9
F_b, F_v, F_c, and $F_{c\perp}$	Dry	1.0	0.8	0.7
	Wet	1.0	0.7	0.5

1. Wet and dry service conditions for sawn lumber, structural glued laminated timber, prefabricated wood I-joists, structural composite lumber, and wood structural panels are specified in 4.1.4, 5.1.5, 7.1.4, 8.1.4, and 9.3.3, respectively.

DESIGN PROVISIONS AND EQUATIONS

3

3.1 General

3.1.1 Scope

Chapter 3 establishes general design provisions that apply to all wood structural members and connections covered under this Specification. Each wood structural member or connection shall be of sufficient size and capacity to carry the applied loads without exceeding the adjusted design values specified herein. Reference design values and specific design provisions applicable to particular wood products or connections are given in other Chapters of this Specification.

3.1.2 Net Section Area

3.1.2.1 The net section area is obtained by deducting from the gross section area the projected area of all material removed by boring, grooving, dapping, notching, or other means. The net section area shall be used in calculating the load carrying capacity of a member, except as specified in 3.6.3 for columns. The effects of any eccentricity of loads applied to the member at the critical net section shall be taken into account.

3.1.2.2 For parallel to grain loading with staggered bolts, drift bolts, drift pins, or lag screws, adjacent fasteners shall be considered as occurring at the same critical section if the parallel to grain spacing between fasteners in adjacent rows is less than four fastener diameters (see Figure 3A).

Figure 3A Spacing of Staggered Fasteners

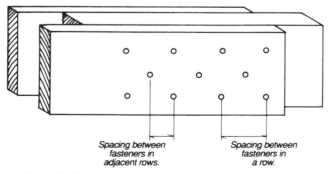

Spacing between fasteners in adjacent rows.

Spacing between fasteners in a row.

3.1.2.3 The net section area at a split ring or shear plate connection shall be determined by deducting from the gross section area the projected areas of the bolt hole and the split ring or shear plate groove within the member (see Figure 3B and Appendix K). Where split ring or shear plate connectors are staggered, adjacent connectors shall be considered as occurring at the same critical section if the parallel to grain spacing between connectors in adjacent rows is less than or equal to one connector diameter (see Figure 3A).

Figure 3B Net Cross Section at a Split Ring or Shear Plate Connection

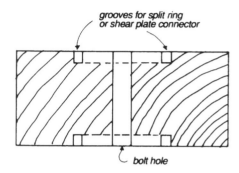

grooves for split ring or shear plate connector

bolt hole

3.1.3 Connections

Structural members and fasteners shall be arranged symmetrically at connections, unless the bending moment induced by an unsymmetrical arrangement (such as lapped joints) has been accounted for in the design. Connections shall be designed and fabricated to insure that each individual member carries its proportional stress.

3.1.4 Time Dependent Deformations

Where members of structural frames are composed of two or more layers or sections, the effect of time dependent deformations shall be accounted for in the design (see 3.5.2 and Appendix F).

3.1.5 Composite Construction

Composite constructions, such as wood-concrete, wood-steel, and wood-wood composites, shall be designed in accordance with principles of engineering mechanics using the adjusted design values for structural members and connections specified herein.

3.2 Bending Members – General

3.2.1 Span of Bending Members

For simple, continuous and cantilevered bending members, the span shall be taken as the distance from face to face of supports, plus ½ the required bearing length at each end.

3.2.2 Lateral Distribution of Concentrated Load

Lateral distribution of concentrated loads from a critically loaded bending member to adjacent parallel bending members by flooring or other cross members shall be permitted to be calculated when determining design bending moment and vertical shear force (see 15.1).

3.2.3 Notches

3.2.3.1 Bending members shall not be notched except as permitted by 4.4.3, 5.4.4, 7.4.4, and 8.4.1. A gradual taper cut from the reduced depth of the member to the full depth of the member in lieu of a square-cornered notch reduces stress concentrations.

3.2.3.2 The stiffness of a bending member, as determined from its cross section, is practically unaffected by a notch with the following dimensions:

notch depth $\leq$ (1/6) (beam depth)
notch length $\leq$ (1/3) (beam depth)

3.2.3.3 See 3.4.3 for effect of notches on shear strength.

3.3 Bending Members – Flexure

3.3.1 Strength in Bending

The actual bending stress or moment shall not exceed the adjusted bending design value.

3.3.2 Flexural Design Equations

3.3.2.1 The actual bending stress induced by a bending moment, M, is calculated as follows:

$$f_b = \frac{Mc}{I} = \frac{M}{S} \qquad (3.3-1)$$

For a rectangular bending member of breadth, b, and depth, d, this becomes:

$$f_b = \frac{M}{S} = \frac{6M}{bd^2} \qquad (3.3-2)$$

3.3.2.2 For solid rectangular bending members with the neutral axis perpendicular to depth at center:

$$I = \frac{bd^3}{12} = \text{moment of inertia} \qquad (3.3-3)$$

$$S = \frac{I}{c} = \frac{bd^2}{6} = \text{section modulus} \qquad (3.3-4)$$

3.3.3 Beam Stability Factor, C_L

3.3.3.1 When the depth of a bending member does not exceed its breadth, $d \leq b$, no lateral support is required and $C_L = 1.0$.

3.3.3.2 When rectangular sawn lumber bending members are laterally supported in accordance with 4.4.1, $C_L = 1.0$.

3.3.3.3 When the compression edge of a bending member is supported throughout its length to prevent lateral displacement, and the ends at points of bearing have lateral support to prevent rotation, $C_L = 1.0$.

3.3.3.4 When the depth of a bending member exceeds its breadth, $d > b$, lateral support shall be provided at points of bearing to prevent rotation and/or lateral displacement at those points. When such lateral support is provided at points of bearing, but no additional lateral support is provided throughout the length of the bending member, the unsupported length, ℓ_u, is the distance between such points of end bearing, or the length of a cantilever. When a bending member is provided with lateral support to prevent rotational and/or lateral displacement at intermediate points as well as at the ends, the unsupported length, ℓ_u, is the distance between such points of intermediate lateral support.

3.3.3.5 The effective span length, ℓ_e, for single span or cantilever bending members shall be determined in accordance with Table 3.3.3.

Table 3.3.3 Effective Length, ℓ_e, for Bending Members

Cantilever[1]	when $\ell_u/d < 7$	when $\ell_u/d \geq 7$
Uniformly distributed load	$\ell_e=1.33\,\ell_u$	$\ell_e=0.90\,\ell_u + 3d$
Concentrated load at unsupported end	$\ell_e=1.87\,\ell_u$	$\ell_e=1.44\,\ell_u + 3d$
Single Span Beam[1,2]	when $\ell_u/d < 7$	when $\ell_u/d \geq 7$
Uniformly distributed load	$\ell_e=2.06\,\ell_u$	$\ell_e=1.63\,\ell_u + 3d$
Concentrated load at center with no intermediate lateral support	$\ell_e=1.80\,\ell_u$	$\ell_e=1.37\,\ell_u + 3d$
Concentrated load at center with lateral support at center	$\ell_e=1.11\,\ell_u$	
Two equal concentrated loads at 1/3 points with lateral support at 1/3 points	$\ell_e=1.68\,\ell_u$	
Three equal concentrated loads at 1/4 points with lateral support at 1/4 points	$\ell_e=1.54\,\ell_u$	
Four equal concentrated loads at 1/5 points with lateral support at 1/5 points	$\ell_e=1.68\,\ell_u$	
Five equal concentrated loads at 1/6 points with lateral support at 1/6 points	$\ell_e=1.73\,\ell_u$	
Six equal concentrated loads at 1/7 points with lateral support at 1/7 points	$\ell_e=1.78\,\ell_u$	
Seven or more equal concentrated loads, evenly spaced, with lateral support at points of load application	$\ell_e=1.84\,\ell_u$	
Equal end moments	$\ell_e=1.84\,\ell_u$	

1. For single span or cantilever bending members with loading conditions not specified in Table 3.3.3:
    $\ell_e = 2.06\,\ell_u$ when $\ell_u/d < 7$
    $\ell_e = 1.63\,\ell_u + 3d$ when $7 \leq \ell_u/d \leq 14.3$
    $\ell_e = 1.84\,\ell_u$ when $\ell_u/d > 14.3$
2. Multiple span applications shall be based on table values or engineering analysis.

3.3.3.6 The slenderness ratio, R_B, for bending members shall be calculated as follows:

$$R_B = \sqrt{\frac{\ell_e d}{b^2}} \qquad (3.3\text{-}5)$$

3.3.3.7 The slenderness ratio for bending members, R_B, shall not exceed 50.

3.3.3.8 The beam stability factor shall be calculated as follows:

$$C_L = \frac{1 + \left(F_{bE}/F_b^*\right)}{1.9} - \sqrt{\left[\frac{1 + \left(F_{bE}/F_b^*\right)}{1.9}\right]^2 - \frac{F_{bE}/F_b^*}{0.95}} \qquad (3.3\text{-}6)$$

where:

F_b^* = reference bending design value multiplied by all applicable adjustment factors except C_{fu}, C_v, and C_L (see 2.3)

$$F_{bE} = \frac{1.20\, E_{min}'}{R_B^2}$$

3.3.3.9 See Appendix D for background information concerning beam stability calculations and Appendix F for information concerning coefficient of variation in modulus of elasticity (COV_E).

3.3.3.10 Members subjected to flexure about both principal axes (biaxial bending) shall be designed in accordance with 3.9.2.

3.4 Bending Members – Shear

3.4.1 Strength in Shear Parallel to Grain (Horizontal Shear)

3.4.1.1 The actual shear stress parallel to grain or shear force at any cross section of the bending member shall not exceed the adjusted shear design value. A check of the strength of wood bending members in shear perpendicular to grain is not required.

3.4.1.2 The shear design procedures specified herein for calculating f_v at or near points of vertical support are limited to solid flexural members such as sawn lumber, structural glued laminated timber, structural composite lumber, or mechanically laminated timber beams. Shear design at supports for built-up components containing load-bearing connections at or near points of support, such as between the web and chord of a truss, shall be based on test or other techniques.

3.4.2 Shear Design Equations

The actual shear stress parallel to grain induced in a sawn lumber, structural glued laminated timber, structural composite lumber, or timber pole or pile bending member shall be calculated as follows:

$$f_v = \frac{VQ}{Ib} \qquad (3.4\text{-}1)$$

For a rectangular bending member of breadth, b, and depth, d, this becomes:

$$f_v = \frac{3V}{2bd} \qquad (3.4\text{-}2)$$

3.4.3 Shear Design

3.4.3.1 When calculating the shear force, V, in bending members:

(a) For beams supported by full bearing on one surface and loads applied to the opposite surface, uniformly distributed loads within a distance from supports equal to the depth of the bending member, d, shall be permitted to be ignored. For beams supported by full bearing on one surface and loads applied to the opposite surface, concentrated loads within a distance, d, from supports shall be permitted to be multiplied by x/d where x is the distance from the beam support face to the load (see Figure 3C).

Figure 3C Shear at Supports

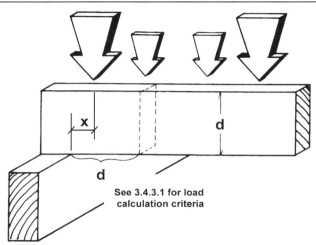

See 3.4.3.1 for load calculation criteria

3 DESIGN PROVISIONS AND EQUATIONS

(b) The largest single moving load shall be placed at a distance from the support equal to the depth of the bending member, keeping other loads in their normal relation and neglecting any load within a distance from a support equal to the depth of the bending member. This condition shall be checked at each support.

(c) With two or more moving loads of about equal weight and in proximity, loads shall be placed in the position that produces the highest shear force, V, neglecting any load within a distance from a support equal to the depth of the bending member.

3.4.3.2 For notched bending members, shear force, V, shall be determined by principles of engineering mechanics (except those given in 3.4.3.1).

(a) For bending members with rectangular cross section and notched on the tension face (see 3.2.3), the adjusted design shear, V_r', shall be calculated as follows:

$$V_r' = \left[\frac{2}{3}F_v'bd_n\right]\left[\frac{d_n}{d}\right]^2 \qquad (3.4-3)$$

where:

    d = depth of unnotched bending member

    d_n = depth of member remaining at a notch

    F_v' = adjusted shear design value parallel to grain

(b) For bending members with circular cross section and notched on the tension face (see 3.2.3), the adjusted design shear, V_r', shall be calculated as follows:

$$V_r' = \left[\frac{2}{3}F_v'A_n\right]\left[\frac{d_n}{d}\right]^2 \qquad (3.4-4)$$

where:

    A_n = cross-sectional area of notched member

(c) For bending members with other than rectangular or circular cross section and notched on the tension face (see 3.2.3), the adjusted design shear, V_r', shall be based on conventional engineering analysis of stress concentrations at notches.

(d) A gradual change in cross section compared with a square notch decreases the actual shear stress parallel to grain nearly to that computed for an unnotched bending member with a depth of d_n.

(e) When a bending member is notched on the compression face at the end as shown in Figure 3D, the adjusted design shear, V_r', shall be calculated as follows:

$$V_r' = \frac{2}{3}F_v'b\left[d-\left(\frac{d-d_n}{d_n}\right)e\right] \qquad (3.4-5)$$

where:

    e = the distance the notch extends inside the inner edge of the support and must be less than or equal to the depth remaining at the notch, $e \le d_n$. If $e > d_n$, d_n shall be used to calculate f_v using Equation 3.4-2.

    d_n = depth of member remaining at a notch meeting the provisions of 3.2.3. If the end of the beam is beveled, as shown by the dashed line in Figure 3D, d_n is measured from the inner edge of the support.

3.4.3.3 When connections in bending members are fastened with split ring connectors, shear plate connectors, bolts, or lag screws (including beams supported by such fasteners or other cases as shown in Figures 3E and 3I) the shear force, V, shall be determined by principles of engineering mechanics (except those given in 3.4.3.1).

Figure 3D Bending Member End-Notched on Compression Face

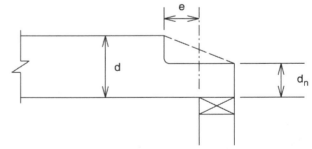

(a) When the connection is less than five times the depth, 5d, of the member from its end, the adjusted design shear, V_r', shall be calculated as follows:

$$V_r' = \left[\frac{2}{3}F_v'bd_e\right]\left[\frac{d_e}{d}\right]^2 \qquad (3.4-6)$$

where:

for split ring or shear plate connections:

d_e = depth of member, less the distance from the unloaded edge of the member to the nearest edge of the nearest split ring or shear plate connector (see Figure 3E).

for bolt or lag screw connections:

d_e = depth of member, less the distance from the unloaded edge of the member to the center of the nearest bolt or lag screw (see Figure 3E)

(b) When the connection is at least five times the depth, 5d, of the member from its end, the adjusted design shear, V_r', shall be calculated as follows:

$$V_r' = \frac{2}{3}F_v'bd_e \qquad (3.4\text{-}7)$$

(c) When concealed hangers are used, the adjusted design shear, V_r', shall be calculated based on the provisions in 3.4.3.2 for notched bending members.

Figure 3E   Effective Depth, d_e, of Members at Connections

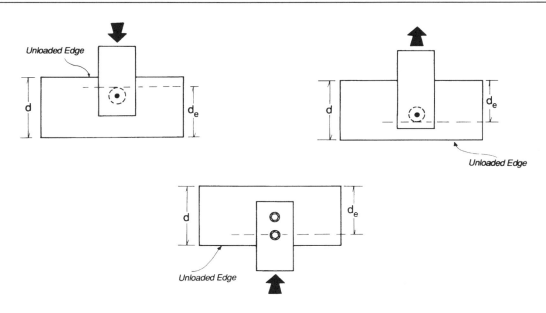

3.5 Bending Members – Deflection

3.5.1 Deflection Calculations

If deflection is a factor in design, it shall be calculated by standard methods of engineering mechanics considering bending deflections and, when applicable, shear deflections. Consideration for shear deflection is required when the reference modulus of elasticity has not been adjusted to include the effects of shear deflection (see Appendix F).

3.5.2 Long-Term Loading

Where total deflection under long-term loading must be limited, increasing member size is one way to

provide extra stiffness to allow for this time dependent deformation (see Appendix F). Total deflection, Δ_T, shall be calculated as follows:

$$\Delta_T = K_{cr}\,\Delta_{LT} + \Delta_{ST}$$

where:

K_{cr} = time dependent deformation (creep) factor

= 1.5 for seasoned lumber, structural glued laminated timber, prefabricated wood I-joists, or structural composite lumber used in dry service conditions as defined in 4.1.4, 5.1.5, 7.1.4, and 8.1.4, respectively.

3

DESIGN PROVISIONS AND EQUATIONS

= 2.0 for structural glued laminated timber used in wet service conditions as defined in 5.1.5.

= 2.0 for wood structural panels used in dry service conditions as defined in 9.1.4.

= 2.0 for unseasoned lumber or for seasoned lumber used in wet service conditions as defined in 4.1.4.

Δ_{LT} = immediate deflection due to the long-term component of the design load

Δ_{ST} = deflection due to the short-term or normal component of the design load

3.6 Compression Members – General

3.6.1 Terminology

For purposes of this Specification, the term "column" refers to all types of compression members, including members forming part of trusses or other structural components.

3.6.2 Column Classifications

3.6.2.1 Simple Solid Wood Columns. Simple columns consist of a single piece or of pieces properly glued together to form a single member (see Figure 3F).

3.6.2.2 Spaced Columns, Connector Joined. Spaced columns are formed of two or more individual members with their longitudinal axes parallel, separated at the ends and middle points of their length by blocking and joined at the ends by split ring or shear plate connectors capable of developing the required shear resistance (see 15.2).

3.6.2.3 Built-Up Columns. Individual laminations of mechanically laminated built-up columns shall be designed in accordance with 3.6.3 and 3.7, except that nailed or bolted built-up columns shall be designed in accordance with 15.3.

3.6.3 Strength in Compression Parallel to Grain

The actual compression stress or force parallel to grain shall not exceed the adjusted compression design value. Calculations of f_c shall be based on the net section area (see 3.1.2) when the reduced section occurs in the critical part of the column length that is most subject to potential buckling. When the reduced section does not occur in the critical part of the column length that is most subject to potential buckling, calculations of f_c shall be based on gross section area. In addition, f_c based on net section area shall not exceed the reference compression design value parallel to grain multiplied

by all applicable adjustment factors except the column stability factor, C_P.

Figure 3F   Simple Solid Column

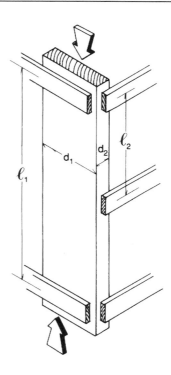

3.6.4 Compression Members Bearing End to End

For end grain bearing of wood on wood, and on metal plates or strips see 3.10.

3.6.5 Eccentric Loading or Combined Stresses

For compression members subject to eccentric loading or combined flexure and axial loading, see 3.9 and 15.4.

3.6.6 Column Bracing

Column bracing shall be installed where necessary to resist wind or other lateral forces (see Appendix A).

3.6.7 Lateral Support of Arches, Studs, and Compression Chords of Trusses

Guidelines for providing lateral support and determining ℓ_e/d in arches, studs, and compression chords of trusses are specified in Appendix A.11.

3.7 Solid Columns

3.7.1 Column Stability Factor, C_P

3.7.1.1 When a compression member is supported throughout its length to prevent lateral displacement in all directions, $C_P = 1.0$.

3.7.1.2 The effective column length, ℓ_e, for a solid column shall be determined in accordance with principles of engineering mechanics. One method for determining effective column length, when end-fixity conditions are known, is to multiply actual column length by the appropriate effective length factor specified in Appendix G, $\ell_e = (K_e)(\ell)$.

3.7.1.3 For solid columns with rectangular cross section, the slenderness ratio, ℓ_e/d, shall be taken as the larger of the ratios ℓ_{e1}/d_1 or ℓ_{e2}/d_2 (see Figure 3F) where each ratio has been adjusted by the appropriate buckling length coefficient, K_e, from Appendix G.

3.7.1.4 The slenderness ratio for solid columns, ℓ_e/d, shall not exceed 50, except that during construction ℓ_e/d shall not exceed 75.

3.7.1.5 The column stability factor shall be calculated as follows:

$$C_P = \frac{1+\left(F_{cE}/F_c^*\right)}{2c} - \sqrt{\left[\frac{1+\left(F_{cE}/F_c^*\right)}{2c}\right]^2 - \frac{F_{cE}/F_c^*}{c}} \qquad (3.7\text{-}1)$$

where:

F_c^* = reference compression design value parallel to grain multiplied by all applicable adjustment factors except C_P (see 2.3)

$$F_{cE} = \frac{0.822\,E_{min}'}{\left(\ell_e/d\right)^2}$$

c = 0.8 for sawn lumber

c = 0.85 for round timber poles and piles

c = 0.9 for structural glued laminated timber or structural composite lumber

3.7.1.6 For especially severe service conditions and/or extraordinary hazard, use of lower adjusted design values may be necessary. See Appendix H for background information concerning column stability calculations and Appendix F for information concerning coefficient of variation in modulus of elasticity (COV_E).

3.7.2 Tapered Columns

For design of a column with rectangular cross section, tapered at one or both ends, the representative dimension, d, for each face of the column shall be derived as follows:

$$d = d_{min} + (d_{max} - d_{min})\left[a - 0.15\left(1 - \frac{d_{min}}{d_{max}}\right)\right] \qquad (3.7\text{-}2)$$

where:

d_{min} = the minimum dimension for that face of the column

d_{max} = the maximum dimension for that face of the column

Support Conditions

Large end fixed, small end unsupported or simply supported	a = 0.70
Small end fixed, large end unsupported or simply supported	a = 0.30
Both ends simply supported:	
Tapered toward one end	a = 0.50
Tapered toward both ends	a = 0.70

For all other support conditions:

$$d = d_{min} + (d_{max} - d_{min})(1/3) \qquad (3.7\text{-}3)$$

Calculations of f_c and C_P shall be based on the representative dimension, d. In addition, f_c at any cross section in the tapered column shall not exceed the reference compression design value parallel to grain multiplied by all applicable adjustment factors except the column stability factor, C_P.

3.7.3 Round Columns

The design of a column of round cross section shall be based on the design calculations for a square column of the same cross-sectional area and having the same degree of taper. Reference design values and special design provisions for round timber poles and piles are provided in Chapter 6.

3.8 Tension Members

3.8.1 Tension Parallel to Grain

The actual tension stress or force parallel to grain shall be based on the net section area (see 3.1.2) and shall not exceed the adjusted tension design value.

3.8.2 Tension Perpendicular to Grain

Designs that induce tension stress perpendicular to grain shall be avoided whenever possible (see References 16 and 19). When tension stress perpendicular to grain cannot be avoided, mechanical reinforcement sufficient to resist all such stresses shall be considered (see References 52 and 53 for additional information).

3.9 Combined Bending and Axial Loading

3.9.1 Bending and Axial Tension

Members subjected to a combination of bending and axial tension (see Figure 3G) shall be so proportioned that:

$$\frac{f_t}{F_t'} + \frac{f_b}{F_b^*} \le 1.0 \qquad (3.9\text{-}1)$$

and

$$\frac{f_b - f_t}{F_b^{**}} \le 1.0 \qquad (3.9\text{-}2)$$

where:

F_b^* = reference bending design value multiplied by all applicable adjustment factors except C_L

F_b^{**} = reference bending design value multiplied by all applicable adjustment factors except C_V

Figure 3G   Combined Bending and Axial Tension

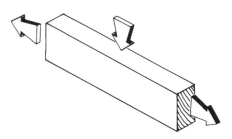

3.9.2 Bending and Axial Compression

Members subjected to a combination of bending about one or both principal axes and axial compression (see Figure 3H) shall be so proportioned that:

$$\left[\frac{f_c}{F_c'}\right]^2 + \frac{f_{b1}}{F_{b1}'\left[1-\left(f_c/F_{cE1}\right)\right]}$$
$$+ \frac{f_{b2}}{F_{b2}'\left[1-\left(f_c/F_{cE2}\right)-\left(f_{b1}/F_{bE}\right)^2\right]} \le 1.0 \qquad (3.9\text{-}3)$$

where:

$$f_c < F_{cE1} = \frac{0.822\,E_{min}{}'}{(\ell_{e1}/d_1)^2}$$ for either uniaxial edge-wise bending or biaxial bending

and

$$f_c < F_{cE2} = \frac{0.822\,E_{min}{}'}{(\ell_{e2}/d_2)^2}$$ for uniaxial flatwise bending or biaxial bending

and

$$f_{b1} < F_{bE} = \frac{1.20\,E_{min}{}'}{(R_B)^2}$$ for biaxial bending

f_{b1} = actual edgewise bending stress (bending load applied to narrow face of member)

f_{b2} = actual flatwise bending stress (bending load applied to wide face of member)

d_1 = wide face dimension (see Figure 3H)

d_2 = narrow face dimension (see Figure 3H)

Effective column lengths, ℓ_{e1} and ℓ_{e2}, shall be determined in accordance with 3.7.1.2. F_c', F_{cE1}, and F_{cE2} shall be determined in accordance with 2.3 and 3.7. F_{b1}', F_{b2}', and F_{bE} shall be determined in accordance with 2.3 and 3.3.3.

3.9.3 Eccentric Compression Loading

See 15.4 for members subjected to combined bending and axial compression due to eccentric loading, or eccentric loading in combination with other loads.

Figure 3H   Combined Bending and Axial Compression

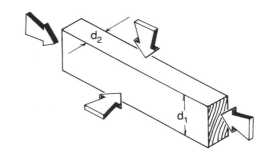

3.10 Design for Bearing

3.10.1 Bearing Parallel to Grain

3.10.1.1 The actual compressive bearing stress parallel to grain shall be based on the net bearing area and shall not exceed the reference compression design value parallel to grain multiplied by all applicable adjustment factors except the column stability factor, C_P.

3.10.1.2 $F_c{}^*$, the reference compression design values parallel to grain multiplied by all applicable adjustment factors except the column stability factor, applies to end-to-end bearing of compression members provided there is adequate lateral support and the end cuts are accurately squared and parallel.

3.10.1.3 When $f_c > (0.75)(F_c{}^*)$ bearing shall be on a metal plate or strap, or on other equivalently durable, rigid, homogeneous material with sufficient stiffness to distribute the applied load. When a rigid insert is required for end-to-end bearing of compression members, it shall be equivalent to 20-gage metal plate or better, inserted with a snug fit between abutting ends.

3.10.2 Bearing Perpendicular to Grain

The actual compression stress perpendicular to grain shall be based on the net bearing area and shall not exceed the adjusted compression design value perpendicular to grain, $f_{c\perp} \leq F_{c\perp}'$. When calculating bearing area at the ends of bending members, no allowance shall be made for the fact that as the member bends, pressure upon the inner edge of the bearing is greater than at the member end.

3.10.3 Bearing at an Angle to Grain

The adjusted bearing design value at an angle to grain (see Figure 3I and Appendix J) shall be calculated as follows:

$$F_\theta' = \frac{F_c^* F_{c\perp}'}{F_c^* \sin^2 \theta + F_{c\perp}' \cos^2 \theta} \qquad (3.10\text{-}1)$$

where:

θ = angle between direction of load and direction of grain (longitudinal axis of member), degrees

3.10.4 Bearing Area Factor, C_b

Reference compression design values perpendicular to grain, $F_{c\perp}$, apply to bearings of any length at the ends of a member, and to all bearings 6" or more in length at any other location. For bearings less than 6" in length and not nearer than 3" to the end of a member, the reference compression design value perpendicular to grain, $F_{c\perp}$, shall be permitted to be multiplied by the following bearing area factor, C_b:

$$C_b = \frac{\ell_b + 0.375}{\ell_b} \qquad (3.10\text{-}2)$$

where:

ℓ_b = bearing length measured parallel to grain, in.

Equation 3.10-2 gives the following bearing area factors, C_b, for the indicated bearing length on such small areas as plates and washers:

Table 3.10.4   Bearing Area Factors, C_b

ℓ_b	0.5"	1"	1.5"	2"	3"	4"	6" or more
C_b	1.75	1.38	1.25	1.19	1.13	1.10	1.00

For round bearing areas such as washers, the bearing length, ℓ_b, shall be equal to the diameter.

Figure 3I   Bearing at an Angle to Grain

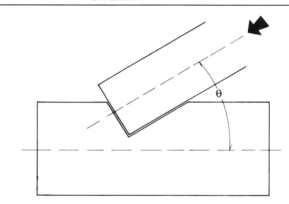

SAWN LUMBER

4

4.1 General

4.1.1 Application

Chapter 4 applies to engineering design with sawn lumber. Design procedures, reference design values and other information herein apply only to lumber complying with the requirements specified below.

4.1.2 Identification of Lumber

4.1.2.1 When the reference design values specified herein are used, the lumber, including end-jointed or edge-glued lumber, shall be identified by the grade mark of, or certificate of inspection issued by, a lumber grading or inspection bureau or agency recognized as being competent (see Reference 31). A distinct grade mark of a recognized lumber grading or inspection bureau or agency, indicating that joint integrity is subject to qualification and quality control, shall be applied to glued lumber products.

4.1.2.2 Lumber shall be specified by commercial species and grade names, or by required levels of design values as listed in Tables 4A, 4B, 4C, 4D, 4E, and 4F (published in the Supplement to this Specification).

4.1.3 Definitions

4.1.3.1 Structural sawn lumber consists of lumber classifications known as "Dimension," "Beams and Stringers," "Posts and Timbers," and "Decking," with design values assigned to each grade.

4.1.3.2 "Dimension" refers to lumber from 2" to 4" (nominal) thick, and 2" (nominal) or more in width. Dimension lumber is further classified as Structural Light Framing, Light Framing, Studs, and Joists and Planks (see References 42, 43, 44, 45, 46, 47, and 49 for additional information).

4.1.3.3 "Beams and Stringers" refers to lumber of rectangular cross section, 5" (nominal) or more thick, with width more than 2" greater than thickness, graded with respect to its strength in bending when loaded on the narrow face.

4.1.3.4 "Posts and Timbers" refers to lumber of square or approximately square cross section, 5" x 5" (nominal) and larger, with width not more than 2" greater than thickness, graded primarily for use as posts or columns carrying longitudinal load.

4.1.3.5 "Decking" refers to lumber from 2" to 4" (nominal) thick, tongued and grooved, or grooved for spline on the narrow face, and intended for use as a roof, floor, or wall membrane. Decking is graded for application in the flatwise direction, with the wide face of the decking in contact with the supporting members, as normally installed.

4.1.4 Moisture Service Condition of Lumber

The reference design values for lumber specified herein are applicable to lumber that will be used under dry service conditions such as in most covered structures, where the moisture content in use will be a maximum of 19%, regardless of the moisture content at the time of manufacture. For lumber used under conditions where the moisture content of the wood in service will exceed 19% for an extended period of time, the design values shall be multiplied by the wet service factors, C_M, specified in Tables 4A, 4B, 4C, 4D, 4E, and 4F.

4.1.5 Lumber Sizes

4.1.5.1 Lumber sizes referred to in this Specification are nominal sizes. Computations to determine the required sizes of members shall be based on the net dimensions (actual sizes) and not the nominal sizes. The dressed sizes specified in Reference 31 shall be accepted as the minimum net sizes associated with nominal dimensions (see Table 1A in the Supplement to this Specification).

4.1.5.2 For 4" (nominal) or thinner lumber, the net DRY dressed sizes shall be used in all computations of structural capacity regardless of the moisture content at the time of manufacture or use.

4.1.5.3 For 5" (nominal) and thicker lumber, the net GREEN dressed sizes shall be used in computations of structural capacity regardless of the moisture content at the time of manufacture or use.

4.1.5.4 Where a design is based on rough sizes or special sizes, the applicable moisture content and size used in design shall be clearly indicated in plans or specifications.

4.1.6 End-Jointed or Edge-Glued Lumber

Reference design values for sawn lumber are applicable to structural end-jointed or edge-glued lumber of the same species and grade. Such use shall include, but not be limited to light framing, studs, joists, planks, and decking. When finger jointed lumber is marked "STUD USE ONLY" or "VERTICAL USE ONLY" such lumber shall be limited to use where any bending or tension stresses are of short duration.

4.1.7 Resawn or Remanufactured Lumber

4.1.7.1 When structural lumber is resawn or re-manufactured, it shall be regraded, and reference design values for the regraded material shall apply (see References 16, 42, 43, 44, 45, 46, 47, and 49).

4.1.7.2 When sawn lumber is cross cut to shorter lengths, the requirements of 4.1.7.1 shall not apply, except for reference bending design values for those Beam and Stringer grades where grading provisions for the middle 1/3 of the length of the piece differ from grading provisions for the outer thirds.

4.2 Reference Design Values

4.2.1 Reference Design Values

Reference design values for visually graded lumber and for mechanically graded dimension lumber are specified in Tables 4A, 4B, 4C, 4D, 4E, and 4F (published in the Supplement to this Specification). The reference design values in Tables 4A, 4B, 4C, 4D, 4E, and 4F are taken from the published grading rules of the agencies cited in References 42, 43, 44, 45, 46, 47, and 49.

4.2.2 Other Species and Grades

Reference design values for species and grades of lumber not otherwise provided herein shall be established in accordance with appropriate ASTM standards and other technically sound criteria (see References 16, 18, 19, and 31).

4.2.3 Basis for Reference Design Values

4.2.3.1 The reference design values in Tables 4A, 4B, 4C, 4D, 4E, and 4F are for the design of structures where an individual member, such as a beam, girder, post or other member, carries or is responsible for carrying its full design load. For repetitive member uses see 4.3.9.

4.2.3.2 Visually Graded Lumber. Reference design values for visually graded lumber in Tables 4A, 4B, 4C, 4D, 4E, and 4F are based on the provisions of ASTM Standards D 245 and D 1990.

4.2.3.3 Machine Stress Rated (MSR) Lumber and Machine Evaluated Lumber (MEL). Reference design values for machine stress rated lumber and machine evaluated lumber in Table 4C are determined by visual grading and nondestructive pretesting of individual pieces.

4.2.4 Modulus of Elasticity, E

4.2.4.1 Average Values. Reference design values for modulus of elasticity assigned to the visually graded species and grades of lumber listed in Tables 4A, 4B, 4C, 4D, 4E, and 4F are average values which conform to ASTM Standards D 245 and D 1990. Adjustments in modulus of elasticity have been taken to reflect increases for seasoning, increases for density where applicable, and, where required, reductions have been made to account for the effect of grade upon stiffness. Reference modulus of elasticity design values are based upon the species or species group average in accordance with ASTM Standards D 1990 and D 2555.

4.2.4.2 Special Uses. Average reference modulus of elasticity design values listed in Tables 4A, 4B, 4C, 4D, 4E, and 4F are to be used in design of repetitive member systems and in calculating the immediate deflection of single members which carry their full design load. In special applications where deflection is a critical factor, or where amount of deformation under long-term loading must be limited, the need for use of a reduced modulus of elasticity design value shall be determined. See Appendix F for provisions on design value adjustments for special end use requirements.

4.2.5 Bending, F_b

4.2.5.1 Dimension Grades. Adjusted bending design values for Dimension grades apply to members with the load applied to either the narrow or wide face.

4.2.5.2 Decking Grades. Adjusted bending design values for Decking grades apply only when the load is applied to the wide face.

4.2.5.3 Post and Timber Grades. Adjusted bending design values for Post and Timber grades apply to members with the load applied to either the narrow or wide face.

4.2.5.4 Beam and Stringer Grades. Adjusted bending design values for Beam and Stringer grades apply to members with the load applied to the narrow face. When Post and Timber sizes of lumber are graded to Beam and Stringer grade requirements, design values for the applicable Beam and Stringer grades shall be used. Such lumber shall be identified in accordance with 4.1.2.1 as conforming to Beam and Stringer grades.

4.2.5.5 Continuous or Cantilevered Beams. When Beams and Stringers are used as continuous or cantilevered beams, the design shall include a requirement that the grading provisions applicable to the middle 1/3 of the length (see References 42, 43, 44, 45, 46, 47, and 49) shall be applied to at least the middle 2/3 of the length of pieces to be used as two span continuous beams, and to the entire length of pieces to be used over three or more spans or as cantilevered beams.

4.2.6 Compression Perpendicular to Grain, $F_{c\perp}$

For sawn lumber, the reference compression design values perpendicular to grain are based on a deformation limit that has been shown by experience to provide for adequate service in typical wood frame construction. The reference compression design values perpendicular to grain specified in Tables 4A, 4B, 4C, 4D, 4E, and 4F are species group average values associated with a deformation level of 0.04" for a steel plate on wood member loading condition. One method for limiting deformation in special applications where it is critical, is use of a reduced compression design value perpendicular to grain. The following equation shall be used to calculate the compression design value perpendicular to grain for a reduced deformation level of 0.02":

$$F_{c\perp 0.02} = 0.73\ F_{c\perp} \qquad (4.2\text{-}1)$$

where:

$F_{c\perp 0.02}$ = compression perpendicular to grain design value at 0.02" deformation limit

$F_{c\perp}$ = reference compression perpendicular to grain design value at 0.04" deformation limit (as published in Tables 4A, 4B, 4C, 4D, 4E, and 4F)

4.3 Adjustment of Reference Design Values

4.3.1 General

Reference design values (F_b, F_t, F_v, $F_{c\perp}$, F_c, E, E_{min}) from Tables 4A, 4B, 4C, 4D, 4E, and 4F shall be multiplied by the adjustment factors specified in Table 4.3.1 to determine adjusted design values (F_b', F_t', F_v', $F_{c\perp}'$, F_c', E', E_{min}').

4.3.2 Load Duration Factor, C_D (ASD only)

All reference design values except modulus of elasticity, E, modulus of elasticity for beam and column stability, E_{min}, and compression perpendicular to grain, $F_{c\perp}$, shall be multiplied by load duration factors, C_D, as specified in 2.3.2.

4.3.3 Wet Service Factor, C_M

Reference design values for structural sawn lumber are based on the moisture service conditions specified in 4.1.4. When the moisture content of structural members in use differs from these moisture service conditions, reference design values shall be multiplied by the wet service factors, C_M, specified in Tables 4A, 4B, 4C, 4D, 4E, and 4F.

4.3.4 Temperature Factor, C_t

When structural members will experience sustained exposure to elevated temperatures up to 150°F (see Appendix C), reference design values shall be multiplied by the temperature factors, C_t, specified in 2.3.3.

planks
v pressure treated

Table 4.3.1 Applicability of Adjustment Factors for Sawn Lumber

		ASD only	ASD and LRFD										LRFD only		
		Load Duration Factor	Wet Service Factor	Temperature Factor	Beam Stability Factor	Size Factor	Flat Use Factor	Incising Factor	Repetitive Member Factor	Column Stability Factor	Buckling Stiffness Factor	Bearing Area Factor	Format Conversion Factor	Resistance Factor	Time Effect Factor
$F_b' = F_b$	x	C_D	C_M	C_t	C_L	C_F	C_{fu}	C_i	C_r	-	-	-	K_F	ϕ_b	λ
$F_t' = F_t$	x	C_D	C_M	C_t	-	C_F	-	C_i	-	-	-	-	K_F	ϕ_t	λ
$F_v' = F_v$	x	C_D	C_M	C_t	-	-	-	C_i	-	-	-	-	K_F	ϕ_v	λ
$F_{c\perp}' = F_{c\perp}$	x	-	C_M	C_t	-	-	-	C_i	-	-	-	C_b	K_F	ϕ_c	λ
$F_c' = F_c$	x	C_D	C_M	C_t	-	C_F	-	C_i	-	C_P	-	-	K_F	ϕ_c	λ
$E' = E$	x	-	C_M	C_t	-	-	-	C_i	-	-	-	-	-	-	-
$E_{min}' = E_{min}$	x	-	C_M	C_t	-	-	-	C_i	-	-	C_T	-	K_F	ϕ_s	-

4

SAWN LUMBER

4.3.5 Beam Stability Factor, C_L

Reference bending design values, F_b, shall be multiplied by the beam stability factor, C_L, specified in 3.3.3.

4.3.6 Size Factor, C_F

4.3.6.1 Reference bending, tension, and compression parallel to grain design values for visually graded dimension lumber 2" to 4" thick shall be multiplied by the size factors specified in Tables 4A and 4B.

4.3.6.2 When the depth of a rectangular sawn lumber bending member 5" or thicker exceeds 12", the reference bending design values, F_b, in Table 4D shall be multiplied by the following size factor:

$$C_F = (12 / d)^{1/9} \leq 1.0 \qquad (4.3\text{-}1)$$

4.3.6.3 For beams of circular cross section with a diameter greater than 13.5", or for 12" or larger square beams loaded in the plane of the diagonal, the size fac-

tor shall be determined in accordance with 4.3.6.2 on the basis of an equivalent conventionally loaded square beam of the same cross-sectional area.

4.3.6.4 Reference bending design values for all species of 2" thick or 3" thick Decking, except Redwood, shall be multiplied by the size factors specified in Table 4E.

4.3.7 Flat Use Factor, C_{fu}

When sawn lumber 2" to 4" thick is loaded on the wide face, multiplying the reference bending design value, F_b, by the flat use factors, C_{fu}, specified in Tables 4A, 4B, 4C, and 4F, shall be permitted.

4.3.8 Incising Factor, C_i

Reference design values shall be multiplied by the following incising factor, C_i, when dimension lumber is incised parallel to grain a maximum depth of 0.4", a maximum length of 3/8", and density of incisions up to

$1100/\text{ft}^2$. Incising factors shall be determined by test or by calculation using reduced section properties for incising patterns exceeding these limits.

Table 4.3.8 Incising Factors, C_i

Design Value	C_i
E, E_{min}	0.95
F_b, F_t, F_c, F_v	0.80
$F_{c\perp}$	1.00

4.3.9 Repetitive Member Factor, C_r

Reference bending design values, F_b, in Tables 4A, 4B, 4C, and 4F for dimension lumber 2" to 4" thick shall be multiplied by the repetitive member factor, C_r = 1.15, when such members are used as joists, truss chords, rafters, studs, planks, decking, or similar members which are in contact or spaced not more than 24" on center, are not less than three in number and are joined by floor, roof or other load distributing elements adequate to support the design load. (A load distributing element is any adequate system that is designed or has been proven by experience to transmit the design load to adjacent members, spaced as described above, without displaying structural weakness or unacceptable deflection. Subflooring, flooring, sheathing, or other covering elements and nail gluing or tongue and groove joints, and through nailing generally meet these criteria.) Reference bending design values in Table 4E for visually graded Decking have already been multiplied by C_r = 1.15.

4.3.10 Column Stability Factor, C_P

Reference compression design values parallel to grain, F_c, shall be multiplied by the column stability factor, C_P, specified in 3.7.

4.3.11 Buckling Stiffness Factor, C_T

Reference modulus of elasticity for beam and column stability, E_{min}, shall be permitted to be multiplied

by the buckling stiffness factor, C_T, as specified in 4.4.2.

4.3.12 Bearing Area Factor, C_b

Reference compression design values perpendicular to grain, $F_{c\perp}$, shall be permitted to be multiplied by the bearing area factor, C_b, as specified in 3.10.4.

4.3.13 Pressure-Preservative Treatment

Reference design values apply to sawn lumber pressure-treated by an approved process and preservative (see Reference 30). Load duration factors greater than 1.6 shall not apply to structural members pressure-treated with water-borne preservatives.

4.3.14 Format Conversion Factor, K_F (LRFD only)

For LRFD, reference design values shall be multiplied by the format conversion factor, K_F, specified in Appendix N.3.1.

4.3.15 Resistance Factor, ϕ (LRFD only)

For LRFD, reference design values shall be multiplied by the resistance factor, ϕ, specified in Appendix N.3.2.

4.3.16 Time Effect Factor, λ (LRFD only)

For LRFD, reference design values shall be multiplied by the time effect factor, λ, specified in Appendix N.3.3.

4.4 Special Design Considerations

4.4.1 Stability of Bending Members

4.4.1.1 Sawn lumber bending members shall be designed in accordance with the lateral stability calculations in 3.3.3 or shall meet the lateral support requirements in 4.4.1.2 and 4.4.1.3.

4.4.1.2 As an alternative to 4.4.1.1, rectangular sawn lumber beams, rafters, joists, or other bending members, shall be designed in accordance with the following provisions to provide restraint against rotation or lateral displacement. If the depth to breadth, d/b, based on nominal dimensions is:

(a) d/b ≤ 2; no lateral support shall be required.

(b) 2 < d/b ≤ 4; the ends shall be held in position, as by full depth solid blocking, bridging, hangers, nailing, or bolting to other framing members, or other acceptable means.

(c) 4 < d/b ≤ 5; the compression edge of the member shall be held in line for its entire length to prevent lateral displacement, as by adequate sheathing or subflooring, and ends at point of bearing shall be held in position to prevent rotation and/or lateral displacement.

(d) 5 < d/b ≤ 6; bridging, full depth solid blocking or diagonal cross bracing shall be installed at intervals not exceeding 8 feet, the compression edge of the member shall be held in line as by adequate sheathing or subflooring, and the ends at points of bearing shall be held in position to prevent rotation and/or lateral displacement.

(e) 6 < d/b ≤ 7; both edges of the member shall be held in line for their entire length and ends at points of bearing shall be held in position to prevent rotation and/or lateral displacement.

4.4.1.3 If a bending member is subjected to both flexure and axial compression, the depth to breadth ratio shall be no more than 5 to 1 if one edge is firmly held in line. If under all combinations of load, the unbraced edge of the member is in tension, the depth to breadth ratio shall be no more than 6 to 1.

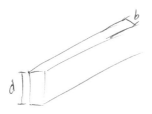

4.4.2 Wood Trusses

4.4.2.1 Increased chord stiffness relative to axial loads when a 2" x 4" or smaller sawn lumber truss compression chord is subjected to combined flexure and axial compression under dry service condition and has 3/8" or thicker plywood sheathing nailed to the narrow face of the chord in accordance with code required roof sheathing fastener schedules (see References 32, 33, and 34), shall be permitted to be accounted for by multiplying the reference modulus of elasticity design value for beam and column stability, E_{min}, by the buckling stiffness factor, C_T, in column stability calculations (see 3.7 and Appendix H). When $\ell_e < 96"$, C_T shall be calculated as follows:

$$C_T = 1 + \frac{K_M \ell_e}{K_T E} \qquad (4.4\text{-}1)$$

where:

ℓ_e = effective column length of truss compression chord (see 3.7)

K_M = 2300 for wood seasoned to 19% moisture content or less at the time of plywood attachment.

= 1200 for unseasoned or partially seasoned wood at the time of plywood attachment.

K_T = 1 − 1.645(COV$_E$)

= 0.59 for visually graded lumber

= 0.75 for machine evaluated lumber (MEL)

= 0.82 for products with COV$_E$ ≤ 0.11 (see Appendix F.2)

When $\ell_e > 96"$, C_T shall be calculated based on ℓ_e = 96".

4.4.2.2 For additional information concerning metal plate connected wood trusses see Reference 9.

4.4.3 Notches

4.4.3.1 End notches, located at the ends of sawn lumber bending members for bearing over a support, shall be permitted, and shall not exceed 1/4 the beam depth (see Figure 4A).

4.4.3.2 Interior notches, located in the outer thirds of the span of a single span sawn lumber bending member, shall be permitted, and shall not exceed 1/6 the depth of the member. Interior notches on the tension side of 3-½" or greater thickness (4" nominal thickness) sawn lumber bending members are not permitted (see Figure 4A).

4.4.3.3 See 3.1.2 and 3.4.3 for effect of notches on strength.

Figure 4A　Notch Limitations for Sawn Lumber Beams

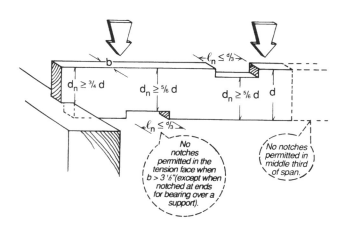

STRUCTURAL GLUED LAMINATED TIMBER

5

5.1 General

5.1.1 Application

5.1.1.1 Chapter 5 applies to engineering design with structural glued laminated timber. Basic requirements are provided in this Specification; for additional detail, see Reference 52.

5.1.1.2 Design procedures, reference design values and other information provided herein apply only to structural glued laminated timber conforming to all pertinent provisions of the specifications referenced in the footnotes to Tables 5A, 5B, 5C, and 5D and produced in accordance with ANSI/AITC A190.1.

5.1.2 Definition

The term "structural glued laminated timber" refers to an engineered, stress rated product of a timber laminating plant, comprising assemblies of specially selected and prepared wood laminations bonded together with adhesives. The grain of all laminations is approximately parallel longitudinally. The separate laminations shall not exceed 2" in net thickness and shall be of:

- one piece, or
- comprised of pieces joined to form any length, or
- pieces placed or glued edge-to-edge to make wider ones, or
- pieces bent to curved-form during gluing.

5.1.3 Standard Sizes

5.1.3.1 Normal standard finished widths of structural glued laminated members shall be as follows:

This Specification is not intended to prohibit other finished widths when required to meet the size requirements of a design or to meet other special requirements.

5.1.3.2 The depth of straight and curved members shall be specified. The length and net dimensions of all members shall also be specified.

5.1.4 Specification

5.1.4.1 For structural glued laminated timber, the following shall be specified:
(a) Dry or wet service conditions.
(b) Laminating combinations or stress requirements.

5.1.4.2 For structural glued laminated hardwood timber, all required reference design values shall be specified for each member.

5.1.5 Service Conditions

5.1.5.1 Reference design values for dry service conditions shall apply when the moisture content in service is less than 16%, as in most covered structures.

5.1.5.2 Reference design values for glued laminated timber shall be multiplied by the wet service factors, C_M, specified in Tables 5A, 5B, 5C, and 5D when the moisture content in service is 16% or greater, as may occur in exterior and submerged construction, or humid environments.

Table 5.1.3 Net Finished Widths of Structural Glued Laminated Timbers

Nominal Width (in.)	3	4	6	8	10	12	14	16
Minimum Net Finished Width (in.)	Western Species							
	2-½	3-⅛	5-⅛	6-¾	8-¾	10-¾	12-¼	14-¼
	Southern Pine							
	-	3	5	6-¾	8-½	10-½	-	-

5.2 Reference Design Values

5.2.1 Reference Design Values

Reference design values for softwood and hardwood structural glued laminated timber are specified in Tables 5A, 5B, 5C, and 5D (published in a separate Supplement to this Specification). The reference design values in Tables 5A, 5B, 5C, and 5D are a compilation of the reference design values provided in the specifications referenced in the footnotes to the tables.

5.2.2 Radial Tension, F_{rt}

For curved bending members, the following reference radial tension design values perpendicular to grain shall apply:

Southern Pine	all loading conditions	$F_{rt} = (1/3)F_v$
Douglas Fir-Larch, Douglas Fir South, Hem-Fir, Western Woods, and Canadian softwood species	wind or earthquake loading	$F_{rt} = (1/3)F_v$
	other types of loading	$F_{rt} = 15$ psi

5.2.3 Other Species and Grades

Reference design values for species and grades of structural glued laminated timber not otherwise provided herein shall be established in accordance with the principles set forth in Reference 22, or shall be based on other substantiated information from an acceptable source.

5.3 Adjustment of Reference Design Values

5.3.1 General

Reference design values (F_b, F_t, F_v, $F_{c\perp}$, F_c, F_{rt}, E, E_{min}) provided in 5.2 and Tables 5A, 5B, 5C, and 5D shall be multiplied by the adjustment factors specified in Table 5.3.1 to determine adjusted design values (F_b', F_t', F_v', $F_{c\perp}'$, F_c', F_{rt}', E', E_{min}').

5.3.2 Load Duration Factor, C_D (ASD only)

All reference design values except modulus of elasticity, E, modulus of elasticity for beam and column stability, E_{min}, and compression perpendicular to grain, $F_{c\perp}$, shall be multiplied by load duration factors, C_D, as specified in 2.3.2.

5.3.3 Wet Service Factor, C_M

Reference design values for structural glued laminated timber are based on the moisture service conditions specified in 5.1.5. When the moisture content of structural members in use differs from these moisture service conditions, reference design values shall be multiplied by the wet service factors, C_M, specified in Tables 5A, 5B, 5C, and 5D.

5

STRUCTURAL GLUED LAMINATED TIMBER

Table 5.3.1 Applicability of Adjustment Factors for Structural Glued Laminated Timber

	ASD only	ASD and LRFD								LRFD only		
	Load Duration Factor	Wet Service Factor	Temperature Factor	Beam Stability Factor[1]	Volume Factor[1]	Flat Use Factor	Curvature Factor	Column Stability Factor	Bearing Area Factor	Format Conversion Factor	Resistance Factor	Time Effect Factor
$F_b' = F_b$ ×	C_D	C_M	C_t	C_L	C_V	C_{fu}	C_c	-	-	K_F	ϕ_b	λ
$F_t' = F_t$ ×	C_D	C_M	C_t	-	-	-	-	-	-	K_F	ϕ_t	λ
$F_v' = F_v$ ×	C_D	C_M	C_t	-	-	-	-	-	-	K_F	ϕ_v	λ
$F_{c\perp}' = F_{c\perp}$ ×	-	C_M	C_t	-	-	-	-	-	C_b	K_F	ϕ_c	λ
$F_c' = F_c$ ×	C_D	C_M	C_t	-	-	-	-	C_P	-	K_F	ϕ_c	λ
$F_{rt}' = F_{rt}$ ×	C_D	C_M	C_t	-	-	-	-	-	-	K_F	ϕ_v	λ
$E' = E$ ×	-	C_M	C_t	-	-	-	-	-	-	-	-	-
$E_{min}' = E_{min}$ ×	-	C_M	C_t	-	-	-	-	-	-	K_F	ϕ_s	-

1. The beam stability factor, C_L, shall not apply simultaneously with the volume factor, C_V, for structural glued laminated timber bending members (see 5.3.6). Therefore, the lesser of these adjustment factors shall apply.

5.3.4 Temperature Factor, C_t

When structural members will experience sustained exposure to elevated temperatures up to 150°F (see Appendix C), reference design values shall be multiplied by the temperature factors, C_t, specified in 2.3.3.

5.3.5 Beam Stability Factor, C_L

Reference bending design values, F_b, shall be multiplied by the beam stability factor, C_L, specified in 3.3.3. The beam stability factor, C_L, shall not apply simultaneously with the volume factor, C_V, for structural glued laminated timber bending members (see 5.3.6). Therefore the lesser of these adjustment factors shall apply.

5.3.6 Volume Factor, C_v

When structural glued laminated timber is loaded perpendicular to the wide face of the laminations, reference bending design values for loading perpendicular to the wide faces of the laminations, F_{bxx}, shall be multiplied by the following volume factor:

$$C_V = \left(\frac{21}{L}\right)^{1/x} \left(\frac{12}{d}\right)^{1/x} \left(\frac{5.125}{b}\right)^{1/x} \le 1.0 \qquad (5.3\text{-}1)$$

where:

L = length of bending member between points of zero moment, ft

d = depth of bending member, in.

b = width (breadth) of bending member. For multiple piece width layups, b = width of widest piece used in the layup. Thus, b ≤ 10.75".

x = 20 for Southern Pine

x = 10 for all other species

The volume factor, C_V, shall not apply simultaneously with the beam stability factor, C_L (see 3.3.3). Therefore, the lesser of these adjustment factors shall apply.

5.3.7 Flat Use Factor, C_{fu}

When structural glued laminated timber is loaded parallel to the wide face of the laminations and the member dimension parallel to the wide face of the laminations is less than 12", multiplying the reference bending design value for loading parallel to the wide faces of the laminations, F_{byy}, by the flat use factors, C_{fu}, specified in Tables 5A, 5B, 5C, and 5D, shall be permitted.

5.3.8 Curvature Factor, C_c

For curved portions of bending members, the reference bending design value shall be multiplied by the following curvature factor:

$$C_c = 1 - (2000)(t / R)^2 \qquad (5.3-2)$$

where:

> t = thickness of lamination, in.
>
> R = radius of curvature of inside face of lamination, in.
>
> t/R ≤ 1/100 for hardwoods and Southern Pine
>
> t/R ≤ 1/125 for other softwoods

The curvature factor shall not apply to reference design values in the straight portion of a member, regardless of curvature elsewhere.

5.3.9 Column Stability Factor, C_P

Reference compression design values parallel to grain, F_c, shall be multiplied by the column stability factor, C_P, specified in 3.7.

5.3.10 Bearing Area Factor, C_b

Reference compression design values perpendicular to grain, $F_{c\perp}$, shall be permitted to be multiplied by the bearing area factor, C_b, as specified in 3.10.4.

5.3.11 Pressure-Preservative Treatment

Reference design values apply to structural glued laminated timber treated by an approved process and preservative (see Reference 30). Load duration factors greater than 1.6 shall not apply to structural members pressure-treated with water-borne preservatives.

5.3.12 Format Conversion Factor, K_F (LRFD only)

For LRFD, reference design values shall be multiplied by the format conversion factor, K_F, specified in Appendix N.3.1.

5.3.13 Resistance Factor, ϕ (LRFD only)

For LRFD, reference design values shall be multiplied by the resistance factor, ϕ, specified in Appendix N.3.2.

5.3.14 Time Effect Factor, λ (LRFD only)

For LRFD, reference design values shall be multiplied by the time effect factor, λ, specified in Appendix N.3.3.

5

STRUCTURAL GLUED LAMINATED TIMBER

5.4 Special Design Considerations

5.4.1 Radial Stress

5.4.1.1 The actual radial stress induced by a bending moment in a curved member of constant rectangular cross section is:

$$f_r = \frac{3M}{2Rbd} \qquad (5.4\text{-}1)$$

where:

> M = bending moment, in.-lbs
>
> R = radius of curvature at center line of member, in.

Curved bending members having a varying rectangular cross section (see Figure 5A) and taper cut structural glued laminated bending members shall be designed in accordance with Reference 52.

Figure 5A   Curved Bending Member

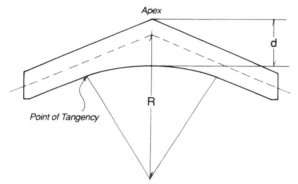

5.4.1.2 When the bending moment is in the direction tending to decrease curvature (increase the radius), the actual radial stress shall not exceed the adjusted radial tension design value perpendicular to grain, $f_r \leq F_{rt}'$, unless mechanical reinforcing sufficient to resist all radial stresses is used (see Reference 52). In no case shall $f_r > (1/3)F_v'$.

5.4.1.3 When the bending moment is in the direction tending to increase curvature (decrease the radius), the actual radial stress shall not exceed the adjusted compression design value perpendicular to grain, $f_r \leq F_{c\perp}'$.

5.4.2 Lateral Stability for Structural Glued Laminated Timber

5.4.2.1 Bending members shall be laterally supported in accordance with 3.3.3, taking into account the provisions of Appendix A.11. The modulus of elasticity of beams loaded parallel to the wide face of the laminations, E_{ymin}, shall be used in beam stability factor calculations.

5.4.2.2 The ratio of tangent point depth to breadth of arches (d/b) shall not exceed 6, based on actual dimensions, when one edge of the arch is braced by decking fastened directly to the arch, or braced at frequent intervals as by girts or roof purlins. When such lateral bracing is not present, d/b shall not exceed 5. Arches shall be designed for lateral stability in accordance with the provisions of 3.7 and 3.9.2.

5.4.3 Deflection

Reference design values for modulus of elasticity in Tables 5A, 5B, 5C, and 5D are average values which include the effects of the grade and placement of laminations used. In special applications where deflection is a critical factor, or where deformation under long-term loading must be limited, the need for use of a reduced reference modulus of elasticity shall be determined. See Appendix F for provisions on design value adjustments for special end use requirements.

5.4.4 Notches

5.4.4.1 The tension side of structural glued laminated timber bending members shall not be notched, except at ends of members for bearing over a support, and notch depth shall not exceed the lesser of 1/10 the depth of the member or 3". The compression side of structural glued laminated timber bending members shall not be notched, except at ends of members, and the notch depth on the compression side shall not exceed 2/5 the depth of the member. Compression side end-notches shall not extend into the middle 1/3 of the span.

> **Exception:** A taper cut on the compression edge at the end of a structural glued laminated timber bending member shall not exceed 2/3 the depth of the member and the length shall not exceed three times the depth of the member, 3d. For tapered beams where the taper extends into the middle 1/3 of the span, special design provisions shall be required.

5.4.4.2 See 3.1.2 and 3.4.3 for effect of notches on strength.

ROUND TIMBER POLES AND PILES

6

6.1 General

6.1.1 Application

6.1.1.1 Chapter 6 applies to engineering design with round timber poles and piles. Design procedures and reference design values herein pertain to the load carrying capacity of poles and piles as structural wood members.

6.1.1.2 This Specification does not apply to the load supporting capacity of the soil.

6.1.2 Specifications

6.1.2.1 The procedures and reference design values herein apply only to timber piles conforming to applicable provisions of ASTM Standard D 25 and only to poles conforming to applicable provisions of ASTM Standard D 3200.

6.1.2.2 Specifications for round timber poles and piles shall include the standard for preservative treatment, pile length, and nominal tip circumference or nominal circumference 3 feet from the butt. Specifications for piles shall state whether piles are to be used as foundation piles, land and fresh water piles, or marine piles.

6.1.3 Standard Sizes

6.1.3.1 Standard sizes for round timber piles are given in ASTM Standard D 25.

6.1.3.2 Standard sizes for round timber poles are given in ASTM Standard D 3200.

6.1.4 Preservative Treatment

6.1.4.1 Reference design values apply to timber poles or piles treated by an approved process and preservative (see Reference 30). Load duration factors greater than 1.6 shall not apply to structural members pressure-treated with water-borne preservatives.

6.1.4.2 Untreated timber poles and piles shall not be used unless the cutoff is below the lowest ground water level expected during the life of the structure, but in no case less than 3 feet below the existing ground water level unless approved by the authority having jurisdiction.

6.2 Reference Design Values

6.2.1 Reference Design Values

6.2.1.1 Reference design values for round timber piles are specified in Table 6A. Reference design values in Table 6A are based on the provisions of ASTM Standard D 2899.

6.2.1.2 Reference design values for round timber poles are specified in Table 6B. Reference design values in Table 6B are based on provisions of ASTM Standard D 3200.

6.2.2 Other Species or Grades

Reference design values for piles of other species or grades shall be determined in accordance with ASTM Standard D 2899.

Table 6A   Reference Design Values for Treated Round Timber Piles

Species	F_c	F_b	F_v	$F_{c\perp}$	E	E_{min}
Reference design values for normal load duration and wet service conditions, psi						
Pacific Coast Douglas Fir[1]	1250	2450	115	230	1,500,000	790,000
Red Oak[2]	1100	2450	135	350	1,250,000	660,000
Red Pine[3]	900	1900	85	155	1,280,000	680,000
Southern Pine[4]	1200	2400	110	250	1,500,000	790,000

1. Pacific Coast Douglas Fir reference design values apply to this species as defined in ASTM Standard D 1760. For connection design use Douglas Fir-Larch reference design values.
2. Red Oak reference design values apply to Northern and Southern Red Oak.
3. Red Pine reference design values apply to Red Pine grown in the United States. For connection design use Northern Pine reference design values.
4. Southern Pine reference design values apply to Loblolly, Longleaf, Shortleaf, and Slash Pines.

Table 6B   Reference Design Values for Poles Graded in Accordance with ASTM D 3200

Species	F_b	F_v	$F_{c\perp}$	F_c	E	E_{min}
Reference design values for normal load duration and wet service conditions, psi						
Pacific Coast Douglas Fir	1850	115	375	1000	1,500,000	790,000
Jack Pine	1500	95	280	800	1,070,000	570,000
Lodgepole Pine	1350	85	240	700	1,080,000	570,000
Northern White Cedar	1050	80	225	525	640,000	340,000
Ponderosa Pine	1300	90	320	650	1,000,000	530,000
Red Pine	1450	85	265	725	1,280,000	680,000
Southern Pine	1700	105	320	900	1,400,000	740,000
Western Hemlock	1650	115	245	900	1,310,000	690,000
Western Larch	2050	120	375	1075	1,460,000	770,000
Western Red Cedar	1350	95	255	750	940,000	500,000

6.3 Adjustment of Reference Design Values

6.3.1 Applicability of Adjustment Factors

Reference design values (F_c, F_b, F_v, $F_{c\perp}$, E, E_{min}) shall be multiplied by all applicable adjustment factors to determine adjusted design values (F_c', F_b', F_v', $F_{c\perp}'$, E', E_{min}'). Table 6.3.1 specifies the adjustment factors which apply to each reference design value for round timber poles and piles.

6.3.2 Load Duration Factor, C_D (ASD only)

All reference design values except modulus of elasticity, E, and modulus of elasticity for beam and column stability, E_{min}, for poles and piles and compression perpendicular to grain $F_{c\perp}$, for poles shall be multiplied

by load duration factors, C_D, as specified in 2.3.2. Load duration factors greater than 1.6 shall not apply to timber poles or piles pressure-treated with water-borne preservatives, (see Reference 30), nor to structural members pressure-treated with fire retardant chemicals (see Table 2.3.2).

6.3.3 Wet Service Factor, C_M

Reference design values apply to wet or dry service conditions.

6.3.4 Temperature Factor, C_t

Reference design values shall be multiplied by temperature factors, C_t, as specified in 2.3.3.

6

ROUND TIMBER POLES AND PILES

Table 6.3.1 Applicability of Adjustment Factors for Round Timber Poles and Piles

		ASD only	ASD and LRFD							LRFD only		
		Load Duration Factor	Temperature Factor	Untreated Factor	Size Factor	Column Stability Factor	Critical Section Factor	Bearing Area Factor	Single Pile Factor	Format Conversion Factor	Resistance Factor	Time Effect Factor
$F_c' = F_c$	x	C_D	C_t	C_u	-	C_P	C_{cs}	-	C_{sp}	K_F	ϕ_c	λ
$F_b' = F_b$	x	C_D	C_t	C_u	C_F	-	-	-	C_{sp}	K_F	ϕ_b	λ
$F_v' = F_v$	x	C_D	C_t	C_u	-	-	-	-	-	K_F	ϕ_v	λ
$F_{c\perp}' = F_{c\perp}$	x	C_D^1	C_t	C_u	-	-	-	C_b	-	K_F	ϕ_c	λ
$E' = E$	x	-	C_t	-	-	-	-	-	-	-	-	-
$E_{min}' = E_{min}$	x	-	C_t	-	-	-	-	-	-	K_F	ϕ_s	-

1. The C_D factor shall not apply to compression perpendicular to grain values for poles.

6.3.5 Untreated Factor, C_u

Reference design values include an adjustment to compensate for strength reduction due to steam conditioning or boultonizing prior to treatment (see Reference 20). Where poles or piles are air dried or kiln dried prior to pressure treatment, or where untreated poles or piles are used, all reference design values except modulus of elasticity, E, and modulus of elasticity for beam and column stability, E_{min}, shall be permitted to be multiplied by the untreated factors, C_u, in Table 6.3.5.

Table 6.3.5 Untreated Factors, C_u, for Timber Poles and Piles

Species	C_u
Pacific Coast Douglas Fir, Red Oak, Red Pine	1.11
Southern Pine	1.18

6.3.6 Beam Stability Factor, C_L

Reference bending design values, F_b, for round timber poles or piles shall not be adjusted for beam stability.

6.3.7 Size Factor, C_F

When pole or pile circumference exceeds 43" (diameter exceeds 13.5") at the critical section in bending, the reference bending design value, F_b, shall be multiplied by the size factor, C_F, specified in 4.3.6.2 and 4.3.6.3.

6.3.8 Column Stability Factor, C_P

Reference compression design values parallel to grain, F_c, for the portion of a timber pole or pile standing unbraced in air, water, or material not capable of lateral support shall be multiplied by the column stability factor, C_P, specified in 3.7.

6.3.9 Critical Section Factor, C_{cs}

Reference compression design values parallel to grain, F_c, for round timber piles are based on the strength at the tip of the pile. Reference compression design values parallel to grain, F_c, for Pacific Coast Douglas Fir and Southern Pine in Table 6A shall be permitted to be increased 0.2% for each foot of length from the tip of the pile to the critical section. The critical section factor, C_{cs}, shall be determined as follows:

$$C_{cs} = 1.0 + (L_c)(0.002) \qquad (6.3\text{-}1)$$

where:

L_c = length from tip of pile to critical section, ft

The increase for location of critical section shall not exceed 10% for any pile ($C_{cs} \leq 1.10$). The critical section factors, C_{cs}, are independent of tapered column provisions in 3.7.2 and both shall be permitted to be used in design calculations.

6.3.10 Bearing Area Factor, C_b

Reference compression design values perpendicular to grain, $F_{c\perp}$, for timber poles or piles shall be permitted to be multiplied by the bearing area factor, C_b, specified in 3.10.4.

6.3.11 Single Pile Factor, C_{sp}

Reference bending design values, F_b, and reference compression design values parallel to grain, F_c, are intended for use when the design encompasses load sharing principles such as occur in a pile cluster. When piles are used in such a manner that each pile supports its own specific load, reference bending design values and reference compression design values parallel to grain shall be multiplied by the single pile factors, C_{sp}, in Table 6.3.11.

Table 6.3.11 Single Pile Factors, C_{sp}, for Round Timber Piles

Reference Design Value	C_{sp}
F_c	0.80
F_b	0.77

6.3.12 Format Conversion Factor, K_F (LRFD only)

For LRFD, reference design values shall be multiplied by the format conversion factor, K_F, specified in Appendix N.3.1.

6.3.13 Resistance Factor, ϕ (LRFD only)

For LRFD, reference design values shall be multiplied by the resistance factor, ϕ, specified in Appendix N.3.2.

6.3.14 Time Effect Factor, λ (LRFD only)

For LRFD, reference design values shall be multiplied by the time effect factor, λ, specified in Appendix N.3.3.

6

ROUND TIMBER POLES AND PILES

PREFABRICATED WOOD I-JOISTS

7

7.1 General

7.1.1 Application

Chapter 7 applies to engineering design with prefabricated wood I-joists. Basic requirements are provided in this Specification. Design procedures and other information provided herein apply only to prefabricated wood I-joists conforming to all pertinent provisions of ASTM D 5055.

7.1.2 Definition

The term "prefabricated wood I-joist" refers to a structural member manufactured using sawn or structural composite lumber flanges and wood structural panel webs bonded together with exterior exposure adhesives, forming an "I" cross-sectional shape.

7.1.3 Identification

When the design procedures and other information provided herein are used, the prefabricated wood I-joists shall be identified with the manufacturer's name and the quality assurance agency's name.

7.1.4 Service Conditions

Reference design values reflect dry service conditions, where the moisture content in service is less than 16%, as in most covered structures. I-joists shall not be used in higher moisture service conditions unless specifically permitted by the I-joist manufacturer.

7.2 Reference Design Values

Reference design values for prefabricated wood I-joists shall be obtained from the prefabricated wood I-joist manufacturer's literature or code evaluation reports.

7.3 Adjustment of Reference Design Values

7.3.1 General

Reference design values (M_r, V_r, R_r, EI, $(EI)_{min}$, K) shall be multiplied by the adjustment factors specified in Table 7.3.1 to determine adjusted design values (M_r', V_r', R_r', EI', $(EI)_{min}'$, K').

7.3.2 Load Duration Factor, C_D (ASD only)

All reference design values except stiffness, EI, $(EI)_{min}$, and K, shall be multiplied by load duration factors, C_D, as specified in 2.3.2.

7.3.3 Wet Service Factor, C_M

Reference design values for prefabricated wood I-joists are applicable to dry service conditions as speci-fied in 7.1.4 where $C_M = 1.0$. When the service conditions differ from the specified conditions, adjustments for high moisture shall be in accordance with information provided by the prefabricated wood I-joist manufacturer.

7.3.4 Temperature Factor, C_t

When structural members will experience sustained exposure to elevated temperatures up to 150°F (see Appendix C), reference design values shall be multiplied by the temperature factors, C_t, specified in 2.3.3. For M_r, V_r, R_r, EI, $(EI)_{min}$, and K use C_t for F_b, F_v, F_v, E, E_{min}, and F_v, respectively.

Table 7.3.1　Applicability of Adjustment Factors for Prefabricated Wood I-Joists

		ASD only	ASD and LRFD				LRFD only		
		Load Duration Factor	Wet Service Factor	Temperature Factor	Beam Stability Factor	Repetitive Member Factor	Format Conversion Factor	Resistance Factor	Time Effect Factor
$M_r' = M_r$	x	C_D	C_M	C_t	C_L	C_r	K_F	ϕ_b	λ
$V_r' = V_r$	x	C_D	C_M	C_t	-	-	K_F	ϕ_v	λ
$R_r' = R_r$	x	C_D	C_M	C_t	-	-	K_F	ϕ_v	λ
$EI' = EI$	x	-	C_M	C_t	-	-	-	-	-
$(EI)_{min}' = (EI)_{min}$	x	-	C_M	C_t	-	-	K_F	ϕ_s	-
$K' = K$	x	-	C_M	C_t	-	-	-	-	-

7.3.5 Beam Stability Factor, C_L

Lateral stability of prefabricated wood I-joists shall be considered. One acceptable method is the procedure of 3.7.1 using the section properties of the compression flange only. The compression flange shall be evaluated as a column continuously restrained in the direction of the web. C_P of the compression flange shall be used as C_L of the joist. Prefabricated wood I-joists shall be restrained against lateral movement and rotation at supports.

7.3.6 Repetitive Member Factor, C_r

For prefabricated wood I-joists with structural composite lumber flanges or sawn lumber flanges, reference moment design resistances shall be multiplied by the repetitive member factor, $C_r = 1.0$.

7.3.7 Pressure-Preservative Treatment

Adjustments to reference design values to account for the effects of pressure-preservative treatment shall be in accordance with information provided by the prefabricated wood I-joist manufacturer.

7.3.8 Format Conversion Factor, K_F (LRFD only)

For LRFD, reference design values shall be multiplied by the format conversion factor, K_F, provided by the wood I-joist manufacturer.

7

PREFABRICATED WOOD I-JOISTS

7.3.9 Resistance Factor, ϕ (LRFD only)

For LRFD, reference design values shall be multiplied by the resistance factor, ϕ, specified in Appendix N.3.2.

7.3.10 Time Effect Factor, λ (LRFD only)

For LRFD, reference design values shall be multiplied by the time effect factor, λ, specified in Appendix N.3.3.

7.4 Special Design Considerations

7.4.1 Bearing

Reference bearing design values, as a function of bearing length, for prefabricated wood I-joists with and without web stiffeners shall be obtained from the prefabricated wood I-joist manufacturer's literature or code evaluation reports.

7.4.2 Load Application

Prefabricated wood I-joists act primarily to resist loads applied to the top flange. Web stiffener requirements, if any, at concentrated loads applied to the top flange and design values to resist concentrated loads applied to the web or bottom flange shall be obtained from the prefabricated wood I-joist manufacturer's literature or code evaluation reports.

7.4.3 Web Holes

The effects of web holes on strength shall be accounted for in the design. Determination of critical shear at a web hole shall consider load combinations of 1.4.4 and partial span loadings defined as live or snow loads applied from each adjacent bearing to the opposite edge of a rectangular hole (centerline of a circular hole). The effects of web holes on deflection are negligible when the number of holes is limited to 3 or less per span. Reference design values for prefabricated wood I-joists with round or rectangular holes shall be obtained from the prefabricated wood I-joist manufacturer's literature or code evaluation reports.

7.4.4 Notches

Notched flanges at or between bearings significantly reduces prefabricated wood I-joist capacity and is beyond the scope of this document. See the manufacturer for more information.

7.4.5 Deflection

Both bending and shear deformations shall be considered in deflection calculations, in accordance with the prefabricated wood I-joist manufacturer's literature or code evaluation reports.

7.4.6 Vertical Load Transfer

I-joists supporting bearing walls located directly above the I-joist support require rim joists, blocking panels, or other means to directly transfer vertical loads from the bearing wall to the supporting structure below.

7.4.7 Shear

Provisions of 3.4.3.1 for calculating shear force, V, shall not be used for design of prefabricated wood I-joist bending members.

STRUCTURAL COMPOSITE LUMBER

8

8.1 General

8.1.1 Application

Chapter 8 applies to engineering design with structural composite lumber. Basic requirements are provided in this Specification. Design procedures and other information provided herein apply only to structural composite lumber conforming to all pertinent provisions of ASTM D5456.

8.1.2 Definitions

8.1.2.1 The term "laminated veneer lumber" refers to a composite of wood veneer sheet elements with wood fiber primarily oriented along the length of the member. Veneer thickness shall not exceed 0.25".

8.1.2.2 The term "parallel strand lumber" refers to a composite of wood strand elements with wood fibers primarily oriented along the length of the member. The least dimension of the strands shall not exceed 0.25" and the average length shall be a minimum of 150 times the least dimension.

8.1.2.3 The term "structural composite lumber" refers to either laminated veneer lumber or parallel strand lumber. These materials are structural members bonded with an exterior adhesive.

8.1.3 Identification

When the design procedures and other information provided herein are used, the structural composite lumber shall be identified with the manufacturer's name and the quality assurance agency's name.

8.1.4 Service Conditions

Reference design values reflect dry service conditions, where the moisture content in service is less than 16%, as in most covered structures. Structural composite lumber shall not be used in higher moisture service conditions unless specifically permitted by the structural composite lumber manufacturer.

8.2 Reference Design Values

Reference design values for structural composite lumber shall be obtained from the structural composite lumber manufacturer's literature or code evaluation report. In special applications where deflection is a critical factor, or where deformation under long-term loading must be limited, the need for use of a reduced modulus of elasticity shall be determined. See Appendix F for provisions on adjusted values for special end use requirements.

8.3 Adjustment of Reference Design Values

8.3.1 General

Reference design values (F_b, F_t, F_v, $F_{c\perp}$, F_c, E, E_{min}) shall be multiplied by the adjustment factors specified in Table 8.3.1 to determine adjusted design values (F_b', F_t', F_v', $F_{c\perp}'$, F_c', E', E_{min}').

Table 8.3.1 Applicability of Adjustment Factors for Structural Composite Lumber

	ASD only	ASD and LRFD							LRFD only		
	Load Duration Factor	Wet Service Factor	Temperature Factor	Beam Stability Factor [1]	Volume Factor [1]	Repetitive Member Factor	Column Stability Factor	Bearing Area Factor	Format Conversion Factor	Resistance Factor	Time Effect Factor
$F_b' = F_b$ x	C_D	C_M	C_t	C_L	C_V	C_r	-	-	K_F	ϕ_b	λ
$F_t' = F_t$ x	C_D	C_M	C_t	-	-	-	-	-	K_F	ϕ_t	λ
$F_v' = F_v$ x	C_D	C_M	C_t	-	-	-	-	-	K_F	ϕ_v	λ
$F_{c\perp}' = F_{c\perp}$ x	-	C_M	C_t	-	-	-	-	C_b	K_F	ϕ_c	λ
$F_c' = F_c$ x	C_D	C_M	C_t	-	-	-	C_P	-	K_F	ϕ_c	λ
$E' = E$ x	-	C_M	C_t	-	-	-	-	-	-	-	-
$E_{min}' = E_{min}$ x	-	C_M	C_t	-	-	-	-	-	K_F	ϕ_s	-

1. See 8.3.6 for information on simultaneous application of the volume factor, C_V, and the beam stability factor, C_L.

8.3.2 Load Duration Factor, C_D (ASD only)

All reference design values except modulus of elasticity, E, modulus of elasticity for beam and column stability, E_{min}, and compression perpendicular to grain, $F_{c\perp}$, shall be multiplied by load duration factors, C_D, as specified in 2.3.2.

8.3.3 Wet Service Factor, C_M

Reference design values for structural composite lumber are applicable to dry service conditions as specified in 8.1.4 where $C_M = 1.0$. When the service conditions differ from the specified conditions, adjustments for high moisture shall be in accordance with information provided by the structural composite lumber manufacturer.

8.3.4 Temperature Factor, C_t

When structural members will experience sustained exposure to elevated temperatures up to 150°F (see Appendix C), reference design values shall be multiplied by the temperature factors, C_t, specified in 2.3.3.

8.3.5 Beam Stability Factor, C_L

Structural composite lumber bending members shall be laterally supported in accordance with 3.3.3.

8.3.6 Volume Factor, C_V

Reference bending design values, F_b, for structural composite lumber shall be multiplied by the volume factor, C_V, and shall be obtained from the structural composite lumber manufacturer's literature or code evaluation reports. When $C_V \leq 1.0$, the volume factor, C_V, shall not apply simultaneously with the beam stability factor, C_L (see 3.3.3) and therefore, the lesser of these adjustment factors shall apply. When $C_V > 1.0$, the volume factor, C_V, shall apply simultaneously with the beam stability factor, C_L (see 3.3.3).

8.3.7 Repetitive Member Factor, C_r

Reference bending design values, F_b, shall be multiplied by the repetitive member factor, $C_r = 1.04$, when such members are used as joists, studs, or similar members which are in contact or spaced not more than 24" on center, are not less than 3 in number and are joined by floor, roof, or other load distributing elements adequate to support the design load. (A load distributing element is any adequate system that is designed or has been proven by experience to transmit the design load to adjacent members, spaced as described above, without displaying structural weakness or unacceptable deflection. Subflooring, flooring, sheathing, or other covering elements and nail gluing or tongue and groove joints, and through nailing generally meet these criteria.)

8.3.8 Column Stability Factor, C_P

Reference compression design values parallel to grain, F_c, shall be multiplied by the column stability factor, C_P, specified in 3.7.

8.3.9 Bearing Area Factor, C_b

Reference compression design values perpendicular to grain, $F_{c\perp}$, shall be permitted to be multiplied by the bearing area factor, C_b, as specified in 3.10.4.

8.3.10 Pressure-Preservative Treatment

Adjustments to reference design values to account for the effects of pressure-preservative treatment shall be in accordance with information provided by the structural composite lumber manufacturer.

8.3.11 Format Conversion Factor, K_F (LRFD only)

For LRFD, reference design values shall be multiplied by the format conversion factor, K_F, specified in Appendix N.3.1.

8.3.12 Resistance Factor, ϕ (LRFD only)

For LRFD, reference design values shall be multiplied by the resistance factor, ϕ, specified in Appendix N.3.2.

8.3.13 Time Effect Factor, λ (LRFD only)

For LRFD, reference design values shall be multiplied by the time effect factor, λ, specified in Appendix N.3.3.

8.4 Special Design Considerations

8.4.1 Notches

8.4.1.1 The tension side of structural composite bending members shall not be notched, except at ends of members for bearing over a support, and notch depth shall not exceed 1/10 the depth of the member. The compression side of structural composite bending members shall not be notched, except at ends of members, and the notch depth on the compression side shall not exceed 2/5 the depth of the member. Compression side end-notches shall not extend into the middle third of the span.

8.4.1.2 See 3.1.2 and 3.4.3 for effect of notches on strength.

WOOD STRUCTURAL PANELS

9

9.1 General

9.1.1 Application

Chapter 9 applies to engineering design with the following wood structural panels: plywood, oriented strand board, and composite panels. Basic requirements are provided in this Specification. Design procedures and other information provided herein apply only to wood structural panels complying with the requirements specified in this Chapter.

9.1.2 Identification

9.1.2.1 When design procedures and other information herein are used, the wood structural panel shall be identified for grade and glue type by the trademarks of an approved testing and grading agency.

9.1.2.2 Wood structural panels shall be specified by span rating, nominal thickness, exposure rating, and grade.

9.1.3 Definitions

9.1.3.1 The term "wood structural panel" refers to a wood-based panel product bonded with a waterproof adhesive. Included under this designation are plywood, oriented strand board (OSB) and composite panels. These panel products meet the requirements of USDOC PS 1 or PS 2 and are intended for structural use in residential, commercial, and industrial applications.

9.1.3.2 The term "composite panel" refers to a wood structural panel comprised of wood veneer and reconstituted wood-based material and bonded with waterproof adhesive.

9.1.3.3 The term "oriented strand board" refers to a mat-formed wood structural panel comprised of thin rectangular wood strands arranged in cross-aligned layers with surface layers normally arranged in the long panel direction and bonded with waterproof adhesive.

9.1.3.4 The term "plywood" refers to a wood structural panel comprised of plies of wood veneer arranged in cross-aligned layers. The plies are bonded with an adhesive that cures on application of heat and pressure.

9.1.4 Service Conditions

9.1.4.1 Reference design values reflect dry service conditions, where the moisture content in service is less than 16%, as in most covered structures.

9.2 Reference Design Values

9.2.1 Panel Stiffness and Strength

9.2.1.1 Reference panel stiffness and strength design values (the product of material and section properties) shall be obtained from an approved source.

9.2.1.2 Due to the orthotropic nature of panels, reference design values shall be provided for the primary and secondary strength axes. The appropriate reference design values shall be applied when designing for each panel orientation. When forces act at an angle to the principal axes of the panel, the capacity of the panel at the angle shall be calculated by adjusting the reference design values for the principal axes using principles of engineering mechanics.

9.2.2 Strength and Elastic Properties

Where required, strength and elastic parameters shall be calculated from reference strength and stiffness design values, respectively, on the basis of tabulated design section properties.

9.2.3 Design Thickness

Nominal thickness shall be used in design calculations. The relationships between span ratings and nominal thicknesses are provided with associated reference design values.

9.2.4 Design Section Properties

Design section properties shall be assigned on the basis of span rating or design thickness and are provided on a per-foot-of-panel-width basis.

9.3 Adjustment of Reference Design Values

9.3.1 General

Reference design values shall be multiplied by the adjustment factors specified in Table 9.3.1 to determine adjusted design values.

9.3.2 Load Duration Factor, C_D (ASD only)

All reference strength design values (F_bS, F_tA, F_vt_v, $F_s(Ib/Q)$, F_cA) shall be multiplied by load duration factors, C_D, as specified in 2.3.2.

9.3.3 Wet Service Factor, C_M, and Temperature Factor, C_t

Reference design values for wood structural panels are applicable to dry service conditions as specified in 9.1.4 where $C_M = 1.0$ and $C_t = 1.0$. When the service conditions differ from the specified conditions, adjustments for high moisture and/or high temperature shall be based on information from an approved source.

Table 9.3.1 Applicability of Adjustment Factors for Wood Structural Panels

	ASD only	ASD and LRFD				LRFD only		
	Load Duration Factor	Wet Service Factor	Temperature Factor	Grade & Construction Factor	Panel Size Factor	Format Conversion Factor	Resistance Factor	Time Effect Factor
$F_bS' = F_bS$ $\times$	C_D	C_M	C_t	C_G	C_s	K_F	ϕ_b	λ
$F_tA' = F_tA$ $\times$	C_D	C_M	C_t	C_G	C_s	K_F	ϕ_t	λ
$F_vt_v' = F_vt_v$ $\times$	C_D	C_M	C_t	C_G	-	K_F	ϕ_v	λ
$F_s(Ib/Q)' = F_s(Ib/Q)$ $\times$	C_D	C_M	C_t	C_G	-	K_F	ϕ_v	λ
$F_cA' = F_cA$ $\times$	C_D	C_M	C_t	C_G	-	K_F	ϕ_c	λ
$EI' = EI$ $\times$	-	C_M	C_t	C_G	-	-	-	-
$EA' = EA$ $\times$	-	C_M	C_t	C_G	-	-	-	-
$G_vt_v' = G_vt_v$ $\times$	-	C_M	C_t	C_G	-	-	-	-
$F_{c\perp}' = F_{c\perp}$ $\times$	-	C_M	C_t	C_G	-	K_F	ϕ_c	λ

9

WOOD STRUCTURAL PANELS

9.3.4 Grade and Construction Factor, C_G, and Panel Size Factor, C_s

Other adjustments to reference panel design values for grade and construction and panel size shall be based on information from an approved source.

9.3.5 Format Conversion Factor, K_F (LRFD only)

For LRFD, reference design values shall be multiplied by the format conversion factor, K_F, specified in Appendix N.3.1.

9.3.6 Resistance Factor, ϕ (LRFD only)

For LRFD, reference design values shall be multiplied by the resistance factor, ϕ, specified in Appendix N.3.2.

9.3.7 Time Effect Factor, λ (LRFD only)

For LRFD, reference design values shall be multiplied by the time effect factor, λ, specified in Appendix N.3.3.

9.4 Design Considerations

9.4.1 Flatwise Bending

Wood structural panels shall be designed for flexure by checking bending moment, shear, and deflection. Adjusted planar shear shall be used as the shear resistance in checking the shear for panels in flatwise bending. Appropriate beam equations shall be used with the design spans as defined below.
  (a) Bending moment-distance between center-line of supports.
  (b) Shear-clear span.
  (c) Deflection-clear span plus the support width factor. For 2" nominal and 4" nominal framing, the support width factor is equal to 0.25" and 0.625", respectively.

9.4.2 Tension in the Plane of the Panel

When wood structural panels are loaded in axial tension, the orientation of the primary strength axis of the panel with respect to the direction of loading, shall be considered in determining adjusted tensile capacity.

9.4.3 Compression in the Plane of the Panel

When wood structural panels are loaded in axial compression, the orientation of the primary strength axis of the panel with respect to the direction of loading, shall be considered in determining the adjusted compressive capacity. In addition, panels shall be designed to prevent buckling.

9.4.4 Planar (Rolling) Shear

The adjusted planar (rolling) shear shall be used in design when the shear force is applied in the plane of wood structural panels.

9.4.5 Through-the-Thickness Shear

The adjusted through-the-thickness shear shall be used in design when the shear force is applied through-the-thickness of wood structural panels.

9.4.6 Bearing

The adjusted bearing design value of wood structural panels shall be used in design when the load is applied perpendicular to the panel face.

MECHANICAL CONNECTIONS

10

10.1 General

10.1.1 Scope

10.1.1.1 Chapter 10 applies to the engineering design of connections using bolts, lag screws, split ring or shear plate connectors, drift bolts, drift pins, wood screws, nails, spikes, timber rivets, metal connector plates or spike grids in sawn lumber, structural glued laminated timber, timber poles, timber piles, structural composite lumber, prefabricated wood I-joists, and wood structural panels. Except where specifically limited elsewhere herein, the provisions of Chapter 10 shall apply to all fastener types covered in Chapters 11, 12, and 13.

10.1.1.2 The requirements of 3.1.3, 3.1.4, and 3.1.5 shall be accounted for in the design of connections.

10.1.1.3 Connection design provisions in Chapters 10, 11, 12, and 13 shall not preclude the use of connections where it is demonstrated by analysis based on generally recognized theory, full-scale, or prototype loading tests, studies of model analogues or extensive experience in use that the connections will perform satisfactorily in their intended end uses (see 1.1.1.3).

10.1.2 Stresses in Members at Connections

Structural members shall be checked for load carrying capacity at connections in accordance with all applicable provisions of this standard including 3.1.2, 3.1.3, and 3.4.3.3. Local stresses in connections using multiple fasteners shall be checked in accordance with principles of engineering mechanics. One method for determining these stresses is provided in Appendix E.

10.1.3 Eccentric Connections

Eccentric connections that induce tension stress perpendicular to grain in the wood shall not be used unless appropriate engineering procedures or tests are employed in the design of such connections to insure that all applied loads will be safely carried by the members and connections. Connections similar to those in Figure 10A are examples of connections requiring appropriate engineering procedures or tests.

10.1.4 Mixed Fastener Connections

Methods of analysis and test data for establishing reference design values for connections made with more than one type of fastener have not been developed. Reference design values and design value adjustments for mixed fastener connections shall be based on tests or other analysis (see 1.1.1.3).

10.1.5 Connection Fabrication

Reference lateral design values for connections in Chapters 11, 12, and 13 are based on:
(a) the assumption that the faces of the members are brought into contact when the fasteners are installed, and
(b) allowance for member shrinkage due to seasonal variations in moisture content (see 10.3.3).

Figure 10A Eccentric Connections

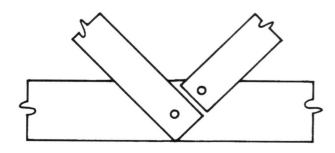
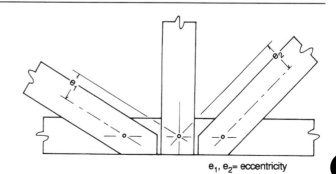

e_1, e_2= eccentricity

10.2 Reference Design Values

10.2.1 Single Fastener Connections

10.2.1.1 Chapters 11, 12, and 13 contain tabulated reference design values and design provisions for calculating reference design values for various types of single fastener connections. Reference design values for connections in a given species apply to all grades of that species unless otherwise indicated. Dowel-type fastener connection reference design values for one species of wood are also applicable to other species having the same or higher dowel bearing strength, F_e.

10.2.1.2 Design provisions and reference design values for dowel-type fastener connections such as bolts, lag screws, wood screws, nails and spikes, drift bolts, and drift pins are provided in Chapter 11.

10.2.1.3 Design provisions and reference design values for split ring and shear plate connections are provided in Chapter 12.

10.2.1.4 Design provisions and reference design values for timber rivet connections are provided in Chapter 13.

10.2.1.5 Wood to wood connections involving spike grids for load transfer shall be designed in accordance with principles of engineering mechanics (see Reference 50 for additional information).

10.2.1.6 Metal plate connected wood truss construction shall be designed in accordance with ANSI/TPI 1.

10.2.2 Multiple Fastener Connections

When a connection contains two or more fasteners of the same type and similar size, each of which exhibits the same yield mode (see Appendix I), the total ad-justed design value for the connection shall be the sum of the adjusted design values for each individual fastener. Local stresses in connections using multiple fasteners shall be evaluated in accordance with principles of engineering mechanics (see 10.1.2).

10.2.3 Design of Metal Parts

Metal plates, hangers, fasteners, and other metal parts shall be designed in accordance with applicable metal design procedures to resist failure in tension, shear, bearing (metal on metal), bending, and buckling (see References 39, 40, and 41). When the capacity of a connection is controlled by metal strength rather than wood strength, metal strength shall not be multiplied by the adjustment factors in this Specification. In addition, metal strength shall not be increased by wind and earthquake factors if design loads have already been reduced by load combination factors (see Reference 5 for additional information).

10.2.4 Design of Concrete or Masonry Parts

Concrete footers, walls, and other concrete or masonry parts shall be designed in accordance with accepted practices (see References 1 and 2). When the capacity of a connection is controlled by concrete or masonry strength rather than wood strength, concrete or masonry strength shall not be multiplied by the adjustment factors in this Specification. In addition, concrete or masonry strength shall not be increased by wind and earthquake factors if design loads have already been reduced by load combination factors (see Reference 5 for additional information).

10.3 Adjustment of Reference Design Values

10.3.1 Applicability of Adjustment Factors

Reference design values (Z, W) shall be multiplied by all applicable adjustment factors to determine ad-justed design values (Z', W'). Table 10.3.1 specifies the adjustment factors which apply to reference lateral design values (Z) and reference withdrawal design values (W) for each fastener type. The actual load applied to a connection shall not exceed the adjusted design value (Z', W') for the connection.

10

MECHANICAL CONNECTIONS

Table 10.3.1 Applicability of Adjustment Factors for Connections

		ASD Only	ASD and LRFD									LRFD Only		
		Load Duration Factor [1]	Wet Service Factor [2]	Temperature Factor	Group Action Factor	Geometry Factor [3]	Penetration Depth Factor [3]	End Grain Factor [3]	Metal Side Plate Factor [3]	Diaphragm Factor [3]	Toe-Nail Factor [3]	Format Conversion Factor	Resistance Factor	Time Effect Factor
Lateral Loads														
Dowel-type Fasteners	$Z' = Z$ x	C_D	C_M	C_t	C_g	C_Δ	-	C_{eg}	-	C_{di}	C_{tn}	K_F	ϕ_z	λ
Split Ring and Shear Plate Connectors	$P' = P$ x	C_D	C_M	C_t	C_g	C_Δ	C_d	-	C_{st}	-	-	K_F	ϕ_z	λ
	$Q' = Q$ x	C_D	C_M	C_t	C_g	C_Δ	C_d	-	-	-	-	K_F	ϕ_z	λ
Timber Rivets	$P' = P$ x	C_D[4]	C_M	C_t	-	-	-	-	C_{st}[5]	-	-	K_F	ϕ_z	λ
	$Q' = Q$ x	C_D[4]	C_M	C_t	-	C_Δ[6]	-	-	C_{st}[5]	-	-	K_F	ϕ_z	λ
Metal Plate Connectors	$Z' = Z$ x	C_D	C_M	C_t	-	-	-	-	-	-	-	K_F	ϕ_z	λ
Spike Grids	$Z' = Z$ x	C_D	C_M	C_t	-	C_Δ	-	-	-	-	-	K_F	ϕ_z	λ
Withdrawal Loads														
Nails, spikes, lag screws, wood screws, and drift pins	$W' = W$ x	C_D	C_M	C_t	-	-	-	C_{eg}	-	-	C_{tn}	K_F	ϕ_z	λ

1. The load duration factor, C_D, shall not exceed 1.6 for connections (see 10.3.2).
2. The wet service factor, C_M, shall not apply to toe-nails loaded in withdrawal (see 11.5.4.1).
3. Specific information concerning geometry factors C_Δ, penetration depth factors C_d, end grain factors C_{eg}, metal side plate factors C_{st}, diaphragm factors C_{di}, and toe-nail factors C_{tn}, is provided in Chapters 11, 12, and 13.
4. The load duration factor, C_D, is only applied when wood capacity (P_w, Q_w) controls (see Chapter 13).
5. The metal side plate factor, C_{st}, is only applied when rivet capacity (P_r, Q_r) controls (see Chapter 13).
6. The geometry factor, C_Δ, is only applied when wood capacity, Q_w, controls (see Chapter 13).

10.3.2 Load Duration Factor, C_D (ASD only)

Reference design values shall be multiplied by the load duration factors, $C_D \leq 1.6$, specified in 2.3.2 and Appendix B, except when the capacity of the connection is controlled by metal strength or strength of concrete/masonry (see 10.2.3, 10.2.4, and Appendix B.3). The impact load duration factor shall not apply to connections.

10.3.3 Wet Service Factor, C_M

Reference design values are for connections in wood seasoned to a moisture content of 19% or less and used under continuously dry conditions, as in most covered structures. For connections in wood that is unsea-soned or partially seasoned, or when connections are exposed to wet service conditions in use, reference design values shall be multiplied by the wet service factors, C_M, specified in Table 10.3.3.

10.3.4 Temperature Factor, C_t

Reference design values shall be multiplied by the temperature factors, C_t, in Table 10.3.4 for connections that will experience sustained exposure to elevated temperatures up to 150°F (see Appendix C).

Table 10.3.3 Wet Service Factors, C_M, for Connections

Fastener Type	Moisture Content		C_M
	At Time of Fabrication	In-Service	
Lateral Loads			
Shear Plates & Split Rings[1]	≤ 19% > 19% any	≤ 19% ≤ 19% > 19%	1.0 0.8 0.7
Metal Connector Plates[2]	≤ 19% > 19% any	≤ 19% ≤ 19% > 19%	1.0 0.8 0.8
Dowel-type Fasteners	≤ 19% > 19% any	≤ 19% ≤ 19% > 19%	1.0 0.4[3] 0.7
Timber Rivets	≤ 19% ≤ 19%	≤ 19% > 19%	1.0 0.8
Withdrawal Loads			
Lag Screws & Wood Screws	any any	≤ 19% > 19%	1.0 0.7
Nails & Spikes	≤ 19% > 19% ≤ 19% > 19%	≤ 19% ≤ 19% > 19% > 19%	1.0 0.25 0.25 1.0
Threaded Hardened Nails	any	any	1.0

1. For split ring or shear plate connectors, moisture content limitations apply to a depth of 3/4" below the surface of the wood.
2. For more information on metal connector plates see Reference 9.
3. C_M = 0.7 for dowel-type fasteners with diameter, D, less than 1/4".
   C_M = 1.0 for dowel-type fastener connections with:
   1) one fastener only, or
   2) two or more fasteners placed in a single row parallel to grain, or
   3) fasteners placed in two or more rows parallel to grain with separate splice plates for each row.

Table 10.3.4 Temperature Factors, C_t, for Connections

In-Service Moisture Conditions[1]	C_t		
	T≤100°F	100°F<T≤125°F	125°F<T≤150°F
Dry	1.0	0.8	0.7
Wet	1.0	0.7	0.5

1. Wet and dry service conditions for connections are specified in 10.3.3.

10.3.5 Fire Retardant Treatment

Adjusted design values for connections in lumber and structural glued laminated timber pressure-treated with fire retardant chemicals shall be obtained from the company providing the treatment and redrying service (see 2.3.4). The impact load duration factor shall not apply to connections in wood pressure-treated with fire retardant chemicals (see Table 2.3.2).

10

MECHANICAL CONNECTIONS

10.3.6 Group Action Factors, C_g

10.3.6.1 Reference lateral design values for split ring connectors, shear plate connectors, or dowel-type fasteners with $D \leq 1''$ in a row shall be multiplied by the following group action factor, C_g:

$$C_g = \left[\frac{m(1-m^{2n})}{n\left[(1+R_{EA}m^n)(1+m)-1+m^{2n}\right]}\right]\left[\frac{1+R_{EA}}{1-m}\right] \quad (10.3\text{-}1)$$

where:

C_g = 1.0 for dowel type fasteners with $D < 1/4''$. Reference design values for timber rivet connections account for group action effects and do not require further modification by the group action factor.

n = number of fasteners in a row

R_{EA} = the lesser of $\dfrac{E_sA_s}{E_mA_m}$ or $\dfrac{E_mA_m}{E_sA_s}$

E_m = modulus of elasticity of main member, psi

E_s = modulus of elasticity of side members, psi

A_m = gross cross-sectional area of main member, in.^{2}

A_s = sum of gross cross-sectional areas of side members, in.^{2}

$m = u - \sqrt{u^2 - 1}$

$u = 1 + \gamma\dfrac{s}{2}\left[\dfrac{1}{E_mA_m} + \dfrac{1}{E_sA_s}\right]$

s = center to center spacing between adjacent fasteners in a row, in.

γ = load/slip modulus for a connection, lbs./in.

= 500,000 lbs./in. for 4" split ring or shear plate connectors

= 400,000 lbs./in. for 2-1/2" split ring or 2-5/8" shear plate connectors

= $(180,000)(D^{1.5})$ for dowel-type fasteners in wood-to-wood connections

= $(270,000)(D^{1.5})$ for dowel-type fasteners in wood-to-metal connections

D = diameter of bolt or lag screw, in.

Group action factors for various connection geometries are provided in Tables 10.3.6A, 10.3.6B, 10.3.6C, and 10.3.6D.

10.3.6.2 For determining group action factors, a row of fasteners is defined as any of the following:

(a) Two or more split rings or shear plate connector units, as defined in 12.1.1, aligned with the direction of load.

(b) Two or more dowel-type fasteners of the same diameter loaded in single or multiple shear and aligned with the direction of load.

When fasteners in adjacent rows are staggered and the distance between adjacent rows is less than 1/4 the distance between the closest fasteners in adjacent rows measured parallel to the rows, the adjacent rows shall be considered as one row for purposes of determining group action factors. For groups of fasteners having an even number of rows, this principle shall apply to each pair of rows. For groups of fasteners having an odd number of rows, the most conservative interpretation shall apply (see Figure 10B).

10.3.6.3 Gross section areas shall be used, with no reductions for net section, when calculating A_m and A_s for determining group action factors. When a member is loaded perpendicular to grain its equivalent cross-sectional area shall be the product of the thickness of the member and the overall width of the fastener group (see Figure 10B). When only one row of fasteners is used, the width of the fastener group shall be the minimum parallel to grain spacing of the fasteners.

Figure 10B Group Action for Staggered Fasteners

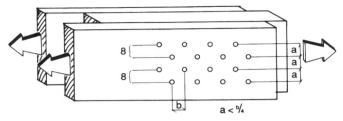

Consider as 2 rows of 8 fasteners

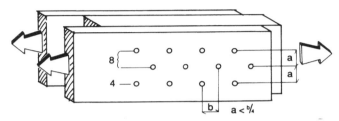

Consider as 1 row of 8 fasteners and 1 row of 4 fasteners

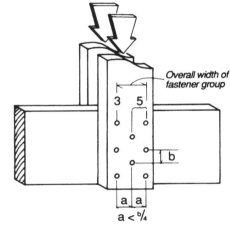

Consider as 1 row of 5 fasteners and 1 row of 3 fasteners

10.3.7 Format Conversion Factor, K_F (LRFD only)

For LRFD, reference design values shall be multiplied by the format conversion factor, K_F, specified in Appendix N.3.1.

10.3.8 Resistance Factor, ϕ (LRFD only)

For LRFD, reference design values shall be multiplied by the resistance factor, ϕ, specified in Appendix N.3.2.

10.3.9 Time Effect Factor, λ (LRFD only)

For LRFD, reference design values shall be multiplied by the time effect factor, λ, specified in Appendix N.3.3.

10

MECHANICAL CONNECTIONS

Table 10.3.6A Group Action Factors, C_g, for Bolt or Lag Screw Connections with Wood Side Members[2]

For D = 1", s = 4", E = 1,400,000 psi

A_s/A_m[1]	A_s[1] in.[2]	Number of fasteners in a row										
		2	3	4	5	6	7	8	9	10	11	12
0.5	5	0.98	0.92	0.84	0.75	0.68	0.61	0.55	0.50	0.45	0.41	0.38
	12	0.99	0.96	0.92	0.87	0.81	0.76	0.70	0.65	0.61	0.57	0.53
	20	0.99	0.98	0.95	0.91	0.87	0.83	0.78	0.74	0.70	0.66	0.62
	28	1.00	0.98	0.96	0.93	0.90	0.87	0.83	0.79	0.76	0.72	0.69
	40	1.00	0.99	0.97	0.95	0.93	0.90	0.87	0.84	0.81	0.78	0.75
	64	1.00	0.99	0.98	0.97	0.95	0.93	0.91	0.89	0.87	0.84	0.82
1	5	1.00	0.97	0.91	0.85	0.78	0.71	0.64	0.59	0.54	0.49	0.45
	12	1.00	0.99	0.96	0.93	0.88	0.84	0.79	0.74	0.70	0.65	0.61
	20	1.00	0.99	0.98	0.95	0.92	0.89	0.86	0.82	0.78	0.75	0.71
	28	1.00	0.99	0.98	0.97	0.94	0.92	0.89	0.86	0.83	0.80	0.77
	40	1.00	1.00	0.99	0.98	0.96	0.94	0.92	0.90	0.87	0.85	0.82
	64	1.00	1.00	0.99	0.98	0.97	0.96	0.95	0.93	0.91	0.90	0.88

1. When A_s/A_m > 1.0, use A_m/A_s and use A_m instead of A_s.
2. Tabulated group action factors (C_g) are conservative for D < 1", s < 4", or E > 1,400,000 psi.

Table 10.3.6B Group Action Factors, C_g, for 4" Split Ring or Shear Plate Connectors with Wood Side Members[2]

s = 9", E = 1,400,000 psi

A_s/A_m[1]	A_s[1] in.[2]	Number of fasteners in a row										
		2	3	4	5	6	7	8	9	10	11	12
0.5	5	0.90	0.73	0.59	0.48	0.41	0.35	0.31	0.27	0.25	0.22	0.20
	12	0.95	0.83	0.71	0.60	0.52	0.45	0.40	0.36	0.32	0.29	0.27
	20	0.97	0.88	0.78	0.69	0.60	0.53	0.47	0.43	0.39	0.35	0.32
	28	0.97	0.91	0.82	0.74	0.66	0.59	0.53	0.48	0.44	0.40	0.37
	40	0.98	0.93	0.86	0.79	0.72	0.65	0.59	0.54	0.49	0.45	0.42
	64	0.99	0.95	0.91	0.85	0.79	0.73	0.67	0.62	0.58	0.54	0.50
1	5	1.00	0.87	0.72	0.59	0.50	0.43	0.38	0.34	0.30	0.28	0.25
	12	1.00	0.93	0.83	0.72	0.63	0.55	0.48	0.43	0.39	0.36	0.33
	20	1.00	0.95	0.88	0.79	0.71	0.63	0.57	0.51	0.46	0.42	0.39
	28	1.00	0.97	0.91	0.83	0.76	0.69	0.62	0.57	0.52	0.47	0.44
	40	1.00	0.98	0.93	0.87	0.81	0.75	0.69	0.63	0.58	0.54	0.50
	64	1.00	0.98	0.95	0.91	0.87	0.82	0.77	0.72	0.67	0.62	0.58

1. When A_s/A_m > 1.0, use A_m/A_s and use A_m instead of A_s.
2. Tabulated group action factors (C_g) are conservative for 2-1/2" split ring connectors, 2-5/8" shear plate connectors, s < 9", or E > 1,400,000 psi.

Table 10.3.6C Group Action Factors, C_g, for Bolt or Lag Screw Connections with Steel Side Plates[1]

A_m/A_s	A_m in.^{2}	Number of fasteners in a row										
		2	3	4	5	6	7	8	9	10	11	12
12	5	0.97	0.89	0.80	0.70	0.62	0.55	0.49	0.44	0.40	0.37	0.34
	8	0.98	0.93	0.85	0.77	0.70	0.63	0.57	0.52	0.47	0.43	0.40
	16	0.99	0.96	0.92	0.86	0.80	0.75	0.69	0.64	0.60	0.55	0.52
	24	0.99	0.97	0.94	0.90	0.85	0.81	0.76	0.71	0.67	0.63	0.59
	40	1.00	0.98	0.96	0.94	0.90	0.87	0.83	0.79	0.76	0.72	0.69
	64	1.00	0.99	0.98	0.96	0.94	0.91	0.88	0.86	0.83	0.80	0.77
	120	1.00	0.99	0.99	0.98	0.96	0.95	0.93	0.91	0.90	0.87	0.85
	200	1.00	1.00	0.99	0.99	0.98	0.97	0.96	0.95	0.93	0.92	0.90
18	5	0.99	0.93	0.85	0.76	0.68	0.61	0.54	0.49	0.44	0.41	0.37
	8	0.99	0.95	0.90	0.83	0.75	0.69	0.62	0.57	0.52	0.48	0.44
	16	1.00	0.98	0.94	0.90	0.85	0.79	0.74	0.69	0.65	0.60	0.56
	24	1.00	0.98	0.96	0.93	0.89	0.85	0.80	0.76	0.72	0.68	0.64
	40	1.00	0.99	0.97	0.95	0.93	0.90	0.87	0.83	0.80	0.77	0.73
	64	1.00	0.99	0.98	0.97	0.95	0.93	0.91	0.89	0.86	0.83	0.81
	120	1.00	1.00	0.99	0.98	0.97	0.96	0.95	0.93	0.92	0.90	0.88
	200	1.00	1.00	0.99	0.99	0.98	0.98	0.97	0.96	0.95	0.94	0.92
24	40	1.00	0.99	0.97	0.95	0.93	0.89	0.86	0.83	0.79	0.76	0.72
	64	1.00	0.99	0.98	0.97	0.95	0.93	0.91	0.88	0.85	0.83	0.80
	120	1.00	1.00	0.99	0.98	0.97	0.96	0.95	0.93	0.91	0.90	0.88
	200	1.00	1.00	0.99	0.99	0.98	0.98	0.97	0.96	0.95	0.93	0.92
30	40	1.00	0.98	0.96	0.93	0.89	0.85	0.81	0.77	0.73	0.69	0.65
	64	1.00	0.99	0.97	0.95	0.93	0.90	0.87	0.83	0.80	0.77	0.73
	120	1.00	0.99	0.99	0.97	0.96	0.94	0.92	0.90	0.88	0.85	0.83
	200	1.00	1.00	0.99	0.98	0.97	0.96	0.95	0.94	0.92	0.90	0.89
35	40	0.99	0.97	0.94	0.91	0.86	0.82	0.77	0.73	0.68	0.64	0.60
	64	1.00	0.98	0.96	0.94	0.91	0.87	0.84	0.80	0.76	0.73	0.69
	120	1.00	0.99	0.98	0.97	0.95	0.92	0.90	0.88	0.85	0.82	0.79
	200	1.00	0.99	0.99	0.98	0.97	0.95	0.94	0.92	0.90	0.88	0.86
42	40	0.99	0.97	0.93	0.88	0.83	0.78	0.73	0.68	0.63	0.59	0.55
	64	0.99	0.98	0.95	0.92	0.88	0.84	0.80	0.76	0.72	0.68	0.64
	120	1.00	0.99	0.97	0.95	0.93	0.90	0.88	0.85	0.81	0.78	0.75
	200	1.00	0.99	0.98	0.97	0.96	0.94	0.92	0.90	0.88	0.85	0.83
50	40	0.99	0.96	0.91	0.85	0.79	0.74	0.68	0.63	0.58	0.54	0.51
	64	0.99	0.97	0.94	0.90	0.85	0.81	0.76	0.72	0.67	0.63	0.59
	120	1.00	0.98	0.97	0.94	0.91	0.88	0.85	0.81	0.78	0.74	0.71
	200	1.00	0.99	0.98	0.96	0.95	0.92	0.90	0.87	0.85	0.82	0.79

For D = 1", s = 4", E_{wood} = 1,400,000 psi, E_{steel} = 30,000,000 psi

1. Tabulated group action factors (C_g) are conservative for D < 1" or s < 4".

10

MECHANICAL CONNECTIONS

Table 10.3.6D Group Action Factors, C_g, for 4" Shear Plate Connectors with Steel Side Plates[1]

A_m/A_s	A_m in.[2]	\multicolumn{11}{c}{s = 9", E_{wood} = 1,400,000 psi, E_{steel} = 30,000,000 psi}
		\multicolumn{11}{c}{Number of fasteners in a row}
		2	3	4	5	6	7	8	9	10	11	12
12	5	0.91	0.75	0.60	0.50	0.42	0.36	0.31	0.28	0.25	0.23	0.21
	8	0.94	0.80	0.67	0.56	0.47	0.41	0.36	0.32	0.29	0.26	0.24
	16	0.96	0.87	0.76	0.66	0.58	0.51	0.45	0.40	0.37	0.33	0.31
	24	0.97	0.90	0.82	0.73	0.64	0.57	0.51	0.46	0.42	0.39	0.35
	40	0.98	0.94	0.87	0.80	0.73	0.66	0.60	0.55	0.50	0.46	0.43
	64	0.99	0.96	0.91	0.86	0.80	0.74	0.69	0.63	0.59	0.55	0.51
	120	0.99	0.98	0.95	0.91	0.87	0.83	0.79	0.74	0.70	0.66	0.63
	200	1.00	0.99	0.97	0.95	0.92	0.89	0.85	0.82	0.79	0.75	0.72
18	5	0.97	0.83	0.68	0.56	0.47	0.41	0.36	0.32	0.28	0.26	0.24
	8	0.98	0.87	0.74	0.62	0.53	0.46	0.40	0.36	0.32	0.30	0.27
	16	0.99	0.92	0.82	0.73	0.64	0.56	0.50	0.45	0.41	0.37	0.34
	24	0.99	0.94	0.87	0.78	0.70	0.63	0.57	0.51	0.47	0.43	0.39
	40	0.99	0.96	0.91	0.85	0.78	0.72	0.66	0.60	0.55	0.51	0.47
	64	1.00	0.97	0.94	0.89	0.84	0.79	0.74	0.69	0.64	0.60	0.56
	120	1.00	0.99	0.97	0.94	0.90	0.87	0.83	0.79	0.75	0.71	0.67
	200	1.00	0.99	0.98	0.96	0.94	0.91	0.89	0.86	0.82	0.79	0.76
24	40	1.00	0.96	0.91	0.84	0.77	0.71	0.65	0.59	0.54	0.50	0.46
	64	1.00	0.98	0.94	0.89	0.84	0.78	0.73	0.68	0.63	0.58	0.54
	120	1.00	0.99	0.96	0.94	0.90	0.86	0.82	0.78	0.74	0.70	0.66
	200	1.00	0.99	0.98	0.96	0.94	0.91	0.88	0.85	0.82	0.78	0.75
30	40	0.99	0.93	0.86	0.78	0.70	0.63	0.57	0.52	0.47	0.43	0.40
	64	0.99	0.96	0.90	0.84	0.78	0.71	0.66	0.60	0.56	0.51	0.48
	120	0.99	0.98	0.94	0.90	0.86	0.81	0.76	0.71	0.67	0.63	0.59
	200	1.00	0.98	0.96	0.94	0.91	0.87	0.83	0.79	0.76	0.72	0.68
35	40	0.98	0.91	0.83	0.74	0.66	0.59	0.53	0.48	0.43	0.40	0.36
	64	0.99	0.94	0.88	0.81	0.73	0.67	0.61	0.56	0.51	0.47	0.43
	120	0.99	0.97	0.93	0.88	0.82	0.77	0.72	0.67	0.62	0.58	0.54
	200	1.00	0.98	0.95	0.92	0.88	0.84	0.80	0.76	0.71	0.68	0.64
42	40	0.97	0.88	0.79	0.69	0.61	0.54	0.48	0.43	0.39	0.36	0.33
	64	0.98	0.92	0.84	0.76	0.69	0.62	0.56	0.51	0.46	0.42	0.39
	120	0.99	0.95	0.90	0.85	0.78	0.72	0.67	0.62	0.57	0.53	0.49
	200	0.99	0.97	0.94	0.90	0.85	0.80	0.76	0.71	0.67	0.62	0.59
50	40	0.95	0.86	0.75	0.65	0.56	0.49	0.44	0.39	0.35	0.32	0.30
	64	0.97	0.90	0.81	0.72	0.64	0.57	0.51	0.46	0.42	0.38	0.35
	120	0.98	0.94	0.88	0.81	0.74	0.68	0.62	0.57	0.52	0.48	0.45
	200	0.99	0.96	0.92	0.87	0.82	0.77	0.71	0.66	0.62	0.58	0.54

1. Tabulated group action factors (C_g) are conservative for 2-5/8" shear plate connectors or s < 9".

DOWEL-TYPE FASTENERS

(BOLTS, LAG SCREWS, WOOD SCREWS, NAILS/SPIKES, DRIFT BOLTS, AND DRIFT PINS)

11

11.1 General

11.1.1 Terminology

11.1.1.1 "Edge distance" is the distance from the edge of a member to the center of the nearest fastener, measured perpendicular to grain. When a member is loaded perpendicular to grain, the loaded edge shall be defined as the edge in the direction toward which the fastener is acting. The unloaded edge shall be defined as the edge opposite the loaded edge (see Figure 11G).

11.1.1.2 "End distance" is the distance measured parallel to grain from the square-cut end of a member to the center of the nearest bolt (see Figure 11G).

11.1.1.3 "Spacing" is the distance between centers of fasteners measured along a line joining their centers (see Figure 11G).

11.1.1.4 A "row of fasteners" is defined as two or more fasteners aligned with the direction of load (see Figure 11G).

11.1.1.5 End distance, edge distance, and spacing requirements herein are based on wood properties. Wood-to-metal and wood-to-concrete connections are subject to placement provisions as shown in 11.5.1, however, applicable end and edge distance and spacing requirements for metal and concrete, also apply (see 10.2.3 and 10.2.4).

11.1.2 Bolts

11.1.2.1 Installation requirements apply to bolts meeting requirements of ANSI/ASME Standard B18.2.1.

11.1.2.2 Holes shall be a minimum of 1/32" to a maximum of 1/16" larger than the bolt diameter. Holes shall be accurately aligned in main members and side plates. Bolts shall not be forcibly driven.

11.1.2.3 A metal plate, metal strap, or washer not less than a standard cut washer shall be between the wood and the bolt head and between the wood and the nut.

11.1.2.4 Edge distance, end distance, and fastener spacing shall not be less than the requirements in Tables 11.5.1A through D.

11.1.3 Lag Screws

11.1.3.1 Installation requirements apply to lag screws meeting requirements of ANSI/ASME Standard B18.2.1. See Appendix L for lag screw dimensions.

11.1.3.2 Lead holes for lag screws loaded laterally and in withdrawal shall be bored as follows to avoid splitting of the wood member during connection fabrication:

(a) The clearance hole for the shank shall have the same diameter as the shank, and the same depth of penetration as the length of unthreaded shank.

(b) The lead hole for the threaded portion shall have a diameter equal to 65% to 85% of the shank diameter in wood with $G > 0.6$, 60% to 75% in wood with $0.5 < G \leq 0.6$, and 40% to 70% in wood with $G \leq 0.5$ (see Table 11.3.2A) and a length equal to at least the length of the threaded portion. The larger percentile in each range shall apply to lag screws of greater diameters.

11.1.3.3 Lead holes or clearance holes shall not be required for 3/8" and smaller diameter lag screws loaded primarily in withdrawal in wood with $G \leq 0.5$ (see Table 11.3.2A), provided that edge distances, end distances, and spacing are sufficient to prevent unusual splitting.

11.1.3.4 The threaded portion of the lag screw shall be inserted in its lead hole by turning with a wrench, not by driving with a hammer.

11.1.3.5 No reduction to reference design values is anticipated if soap or other lubricant is used on the lag screw or in the lead holes to facilitate insertion and to prevent damage to the lag screw.

11.1.3.6 Minimum penetration (not including the length of the tapered tip) of the lag screw into the main member for single shear connections or the side member for double shear connections shall be four times the diameter, $p_{min} = 4D$.

11.1.3.7 Edge distance, end distance, and fastener spacing shall not be less than the requirements in Tables 11.5.1A through E.

11.1.4 Wood Screws

11.1.4.1 Installation requirements apply to wood screws meeting requirements of ANSI/ASME Standard B18.6.1.

11.1.4.2 Lead holes for wood screws loaded in withdrawal shall have a diameter equal to approximately 90% of the wood screw root diameter in wood with $G > 0.6$, and approximately 70% of the wood screw root diameter in wood with $0.5 < G \leq 0.6$. Wood with $G \leq 0.5$ (see Table 11.3.2A) is not required to have a lead hole for insertion of wood screws.

11.1.4.3 Lead holes for wood screws loaded laterally shall be bored as follows:

(a) For wood with G > 0.6 (see Table 11.3.2A), the part of the lead hole receiving the shank shall have about the same diameter as the shank, and that receiving the threaded portion shall have about the same diameter as the screw at the root of the thread (see Reference 8).

(b) For G ≤ 0.6 (see Table 11.3.2A), the part of the lead hole receiving the shank shall be about 7/8 the diameter of the shank and that receiving the threaded portion shall be about 7/8 the diameter of the screw at the root of the thread (see Reference 8).

11.1.4.4 The wood screw shall be inserted in its lead hole by turning with a screw driver or other tool, not by driving with a hammer.

11.1.4.5 No reduction to reference design values is anticipated if soap or other lubricant is used on the wood screw or in the lead holes to facilitate insertion and to prevent damage to the wood screw.

11.1.4.6 Minimum penetration of the wood screw into the main member for single shear connections or the side member for double shear connections shall be six times the diameter, $p_{min} = 6D$.

11.1.4.7 Edge distances, end distances, and spacings shall be sufficient to prevent splitting of the wood.

11.1.5 Nails and Spikes

11.1.5.1 Installation requirements apply to common steel wire nails and spikes, box nails, and threaded hardened-steel nails meeting requirements in ASTM F1667. Nail specifications for engineered construction shall include the minimum lengths and diameters for the nails and spikes to be used.

11.1.5.2 Threaded, hardened-steel nails, and spikes shall be made of high carbon steel wire, headed, pointed, annularly or helically threaded, and heat-treated and tempered to provide greater yield strength than for common wire nails of corresponding size.

11.1.5.3 Reference design values herein apply to nailed and spiked connections either with or without bored holes. When a bored hole is desired to prevent splitting of wood, the diameter of the bored hole shall not exceed 90% of the nail or spike diameter for wood with G > 0.6, nor 75% of the nail or spike diameter for wood with G ≤ 0.6 (see Table 11.3.2A).

11.1.5.4 Toe-nails shall be driven at an angle of approximately 30° with the member and started approximately 1/3 the length of the nail from the member end (see Figure 11A).

Figure 11A   Toe-Nail Connection

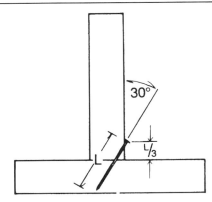

11.1.5.5 Minimum penetration of the nail or spike into the main member for single shear connections or the side member for double shear connections shall be six times the diameter, $p_{min} = 6D$.

Exception: Symmetric double shear connections when 12d or smaller nails extend at least three diameters beyond the side member and are clinched, and side members are at least 3/8" thick.

11.1.5.6 Edge distances, end distances, and spacings shall be sufficient to prevent splitting of the wood.

11.1.6 Drift Bolts and Drift Pins

11.1.6.1 Lead holes shall be drilled 0" to 1/32" smaller than the actual pin diameter.

11.1.6.2 Additional penetration of pin into members shall be provided in lieu of the washer, head, and nut on a common bolt (see Reference 53 for additional information).

11.1.6.3 Edge distance, end distance, and fastener spacing shall not be less than the requirements in Tables 11.5.1A through D.

11.1.7 Other Dowel-Type Fasteners

When fastener type or connection fabrication and assembly requirements vary from those specified in 11.1.2, 11.1.3, 11.1.4, 11.1.5, and 11.1.6, provisions of 11.3 shall be permitted to be used in calculation of reference lateral design values provided allowance is made to account for such variation. Edge distances, end distances, and spacings shall be sufficient to prevent splitting of the wood.

DOWEL-TYPE FASTENERS

11

11.2 Reference Withdrawal Design Values

11.2.1 Lag Screws

11.2.1.1 The reference withdrawal design values, in lb/in. of penetration, for a single lag screw inserted in side grain, with the lag screw axis perpendicular to the wood fibers, shall be determined from Table 11.2A or Equation 11.2-1, within the range of specific gravities and screw diameters given in Table 11.2A. Reference withdrawal design values, W, shall be multiplied by all applicable adjustment factors (see Table 10.3.1) to obtain adjusted withdrawal design values, W'.

$$W = 1800\ G^{3/2}D^{3/4} \qquad\qquad (11.2\text{-}1)$$

11.2.1.2 When lag screws are loaded in withdrawal from end grain, reference withdrawal design values, W, shall be multiplied by the end grain factor, $C_{eg} = 0.75$.

11.2.1.3 When lag screws are loaded in withdrawal, the tensile strength of the lag screw at the net (root) section shall not be exceeded (see 10.2.3).

Table 11.2A   Lag Screw Reference Withdrawal Design Values (W)[1]

Tabulated withdrawal design values (W) are in pounds per inch of thread penetration into side grain of main member. Length of thread penetration in main member shall not include the length of the tapered tip (see Appendix L).

Specific Gravity, G	Lag Screw Unthreaded Shank Diameter, D										
	1/4"	5/16"	3/8"	7/16"	1/2"	5/8"	3/4"	7/8"	1"	1-1/8"	1-1/4"
0.73	397	469	538	604	668	789	905	1016	1123	1226	1327
0.71	381	450	516	579	640	757	868	974	1077	1176	1273
0.68	357	422	484	543	600	709	813	913	1009	1103	1193
0.67	349	413	473	531	587	694	796	893	987	1078	1167
0.58	281	332	381	428	473	559	641	719	795	869	940
0.55	260	307	352	395	437	516	592	664	734	802	868
0.51	232	274	314	353	390	461	528	593	656	716	775
0.50	225	266	305	342	378	447	513	576	636	695	752
0.49	218	258	296	332	367	434	498	559	617	674	730
0.47	205	242	278	312	345	408	467	525	580	634	686
0.46	199	235	269	302	334	395	453	508	562	613	664
0.44	186	220	252	283	312	369	423	475	525	574	621
0.43	179	212	243	273	302	357	409	459	508	554	600
0.42	173	205	235	264	291	344	395	443	490	535	579
0.41	167	198	226	254	281	332	381	428	473	516	559
0.40	161	190	218	245	271	320	367	412	455	497	538
0.39	155	183	210	236	261	308	353	397	438	479	518
0.38	149	176	202	227	251	296	340	381	422	461	498
0.37	143	169	194	218	241	285	326	367	405	443	479
0.36	137	163	186	209	231	273	313	352	389	425	460
0.35	132	156	179	200	222	262	300	337	373	407	441
0.31	110	130	149	167	185	218	250	281	311	339	367

1. Tabulated withdrawal design values (W) for lag screw connections shall be multiplied by all applicable adjustment factors (see Table 10.3.1).

11.2.2 Wood Screws

11.2.2.1 The reference withdrawal design value, in lb/in. of penetration, for a single wood screw (cut thread or rolled thread) inserted in side grain, with the wood screw axis perpendicular to the wood fibers, shall be determined from Table 11.2B or Equation 11.2-2, within the range of specific gravities and screw diameters given in Table 11.2B. Reference withdrawal design values, W, shall be multiplied by all applicable adjustment factors (see Table 10.3.1) to obtain adjusted withdrawal design values, W'.

$$W = 2850 \, G^2 D \qquad\qquad (11.2\text{-}2)$$

11.2.2.2 Wood screws shall not be loaded in withdrawal from end grain of wood.

11.2.2.3 When wood screws are loaded in withdrawal, the adjusted tensile strength of the wood screw at net (root) section shall not be exceeded (see 10.2.3).

11.2.3 Nails and Spikes

11.2.3.1 The reference withdrawal design value, in lb/in. of penetration, for a single nail or spike driven in the side grain of the main member, with the nail or spike axis perpendicular to the wood fibers, shall be determined from Table 11.2C or Equation 11.2-3, within the range of specific gravities and nail or spike diameters given in Table 11.2C. Reference withdrawal design values, W, shall be multiplied by all applicable adjustment factors (see Table 10.3.1) to obtain adjusted withdrawal design values, W'.

$$W = 1380 \, G^{5/2} D \qquad\qquad (11.2\text{-}3)$$

11.2.3.2 Nails and spikes shall not be loaded in withdrawal from end grain of wood.

11.2.4 Drift Bolts and Drift Pins

Drift bolt and drift pin connections loaded in withdrawal shall be designed in accordance with principles of engineering mechanics.

DOWEL-TYPE FASTENERS

11

Table 11.2B Cut Thread or Rolled Thread Wood Screw Reference Withdrawal Design Values (W)[1]

Tabulated withdrawal design values (W) are in pounds per inch of thread penetration into side grain of main member. Thread length is approximately 2/3 the total wood screw length (see Appendix L).

Specific Gravity, G	Wood Screw Number										
	6	7	8	9	10	12	14	16	18	20	24
0.73	209	229	249	268	288	327	367	406	446	485	564
0.71	198	216	235	254	272	310	347	384	421	459	533
0.68	181	199	216	233	250	284	318	352	387	421	489
0.67	176	193	209	226	243	276	309	342	375	409	475
0.58	132	144	157	169	182	207	232	256	281	306	356
0.55	119	130	141	152	163	186	208	231	253	275	320
0.51	102	112	121	131	141	160	179	198	217	237	275
0.50	98	107	117	126	135	154	172	191	209	228	264
0.49	94	103	112	121	130	147	165	183	201	219	254
0.47	87	95	103	111	119	136	152	168	185	201	234
0.46	83	91	99	107	114	130	146	161	177	193	224
0.44	76	83	90	97	105	119	133	148	162	176	205
0.43	73	79	86	93	100	114	127	141	155	168	196
0.42	69	76	82	89	95	108	121	134	147	161	187
0.41	66	72	78	85	91	103	116	128	141	153	178
0.40	63	69	75	81	86	98	110	122	134	146	169
0.39	60	65	71	77	82	93	105	116	127	138	161
0.38	57	62	67	73	78	89	99	110	121	131	153
0.37	54	59	64	69	74	84	94	104	114	125	145
0.36	51	56	60	65	70	80	89	99	108	118	137
0.35	48	53	57	62	66	75	84	93	102	111	130
0.31	38	41	45	48	52	59	66	73	80	87	102

1. Tabulated withdrawal design values (W) for wood screw connections shall be multiplied by all applicable adjustment factors (see Table 10.3.1).

Table 11.2C Nail and Spike Reference Withdrawal Design Values (W)[1]

Tabulated withdrawal design values (W) are in pounds per inch of penetration into side grain of main member (see Appendix L).

Specific Gravity, G	Common Wire Nails, Box Nails, and Common Wire Spikes — Diameter, D															Threaded Nails — Wire Diameter, D				
	0.099"	0.113"	0.128"	0.131"	0.135"	0.148"	0.162"	0.192"	0.207"	0.225"	0.244"	0.263"	0.283"	0.312"	0.375"	0.120"	0.135"	0.148"	0.177"	0.207"
0.73	62	71	80	82	85	93	102	121	130	141	153	165	178	196	236	82	93	102	121	141
0.71	58	66	75	77	79	87	95	113	121	132	143	154	166	183	220	77	87	95	113	132
0.68	52	59	67	69	71	78	85	101	109	118	128	138	149	164	197	69	78	85	101	118
0.67	50	57	65	66	68	75	82	97	105	114	124	133	144	158	190	66	75	82	97	114
0.58	35	40	45	46	48	52	57	68	73	80	86	93	100	110	133	46	52	57	68	80
0.55	31	35	40	41	42	46	50	59	64	70	76	81	88	97	116	41	46	50	59	70
0.51	25	29	33	34	35	38	42	49	53	58	63	67	73	80	96	34	38	42	49	58
0.50	24	28	31	32	33	36	40	47	50	55	60	64	69	76	91	32	36	40	47	55
0.49	23	26	30	30	31	34	38	45	48	52	57	61	66	72	87	30	34	38	45	52
0.47	21	24	27	27	28	31	34	40	43	47	51	55	59	65	78	27	31	34	40	47
0.46	20	22	25	26	27	29	32	38	41	45	48	52	56	62	74	26	29	32	38	45
0.44	18	20	23	23	24	26	29	34	37	40	43	47	50	55	66	23	26	29	34	40
0.43	17	19	21	22	23	25	27	32	35	38	41	44	47	52	63	22	25	27	32	38
0.42	16	18	20	21	21	23	26	30	33	35	38	41	45	49	59	21	23	26	30	35
0.41	15	17	19	19	20	22	24	29	31	33	36	39	42	46	56	19	22	24	29	33
0.40	14	16	18	18	19	21	23	27	29	31	34	37	40	44	52	18	21	23	27	31
0.39	13	15	17	17	18	19	21	25	27	29	32	34	37	41	49	17	19	21	25	29
0.38	12	14	16	16	17	18	20	24	25	28	30	32	35	38	46	16	18	20	24	28
0.37	11	13	15	15	16	17	19	22	24	26	28	30	33	36	43	15	17	19	22	26
0.36	11	12	14	14	14	16	17	21	22	24	26	28	30	33	40	14	16	17	21	24
0.35	10	11	13	13	14	15	16	19	21	23	24	26	28	31	38	13	15	16	19	23
0.31	7	8	9	10	10	11	12	14	15	17	18	19	21	23	28	10	11	12	14	17

1. Tabulated withdrawal design values (W) for nail or spike connections shall be multiplied by all applicable adjustment factors (see Table 10.3.1).

11.3 Reference Lateral Design Values

11.3.1 Yield Limit Equations

For single shear and symmetric double shear connections using dowel-type fasteners (see Appendix I, Figures 11B and 11C) where:

(a) the faces of the connected members are in contact

(b) the load acts perpendicular to the axis of the dowel

(c) edge distances, end distances, and spacing are not less than the requirements in 11.5, and

(d) the depth of fastener penetration in the main member for single shear connections or the side member holding the point for double shear connections is greater than or equal to the minimum penetration required (see 11.1).

The reference design value, Z, shall be the minimum computed yield mode value using equations in Tables 11.3.1A and B. Reference design values for connections with bolts (see Tables 11A through I), lag screws (see Tables 11J and K), wood screws (see Tables 11L and M), and nails and spikes (see Tables 11N through R) are calculated for common connection conditions in accordance with yield mode equations in Tables 11.3.1A and B.

Table 11.3.1A   Yield Limit Equations

Yield Mode	Single Shear		Double Shear	
I_m	$Z = \dfrac{D\,\ell_m\,F_{em}}{R_d}$	(11.3-1)	$Z = \dfrac{D\,\ell_m\,F_{em}}{R_d}$	(11.3-7)
I_s	$Z = \dfrac{D\,\ell_s\,F_{es}}{R_d}$	(11.3-2)	$Z = \dfrac{2\,D\,\ell_s\,F_{es}}{R_d}$	(11.3-8)
II	$Z = \dfrac{k_1\,D\,\ell_s\,F_{es}}{R_d}$	(11.3-3)		
III_m	$Z = \dfrac{k_2\,D\,\ell_m\,F_{em}}{(1+2R_e)\,R_d}$	(11.3-4)		
III_s	$Z = \dfrac{k_3\,D\,\ell_s\,F_{em}}{(2+R_e)\,R_d}$	(11.3-5)	$Z = \dfrac{2\,k_3\,D\,\ell_s\,F_{em}}{(2+R_e)\,R_d}$	(11.3-9)
IV	$Z = \dfrac{D^2}{R_d}\sqrt{\dfrac{2\,F_{em}\,F_{yb}}{3\,(1+R_e)}}$	(11.3-6)	$Z = \dfrac{2\,D^2}{R_d}\sqrt{\dfrac{2\,F_{em}\,F_{yb}}{3\,(1+R_e)}}$	(11.3-10)

Notes:

$$k_1 = \frac{\sqrt{R_e + 2R_e^2(1+R_t+R_t^2)+R_t^2 R_e^3}-R_e(1+R_t)}{(1+R_e)}$$

$$k_2 = -1+\sqrt{2(1+R_e)+\frac{2F_{yb}(1+2R_e)D^2}{3F_{em}\ell_m^2}}$$

$$k_3 = -1+\sqrt{\frac{2(1+R_e)}{R_e}+\frac{2F_{yb}(2+R_e)D^2}{3F_{em}\ell_s^2}}$$

D = diameter, in. (see 11.3.6)
F_{yb} = dowel bending yield strength, psi
R_d = reduction term (see Table 11.3.1B)
R_e = F_{em}/F_{es}
R_t = ℓ_m/ℓ_s
ℓ_m = main member dowel bearing length, in.
ℓ_s = side member dowel bearing length, in.
F_{em} = main member dowel bearing strength, psi (see Table 11.3.2)
F_{es} = side member dowel bearing strength, psi (see Table 11.3.2)

Table 11.3.1B Reduction Term, R_d

Fastener Size	Yield Mode	Reduction Term, R_d
$0.25" \leq D \leq 1"$	I_m, I_s	$4 K_\theta$
	II	$3.6 K_\theta$
	III_m, III_s, IV	$3.2 K_\theta$
$D < 0.25"$	I_m, I_s, II, III_m, III_s, IV	K_D [1]

Notes:

K_θ = $1 + 0.25(\theta/90)$

θ = maximum angle of load to grain ($0° \leq \theta \leq 90°$) for any member in a connection

D = diameter, in. (see 11.3.6)

K_D = 2.2 for $D \leq 0.17"$

K_D = $10D + 0.5$ for $0.17" < D < 0.25"$

1. For threaded fasteners where nominal diameter (see Appendix L) is greater than or equal to 0.25" and root diameter is less than 0.25", $R_d = K_D K_\theta$.

11.3.2 Dowel Bearing Strength

11.3.2.1 Dowel bearing strengths, F_e, for parallel or perpendicular to grain loading are provided for dowel-type fasteners with $1/4" \leq D \leq 1"$ in Table 11.3.2. When fastener diameter, $D < 1/4"$, a single dowel bearing strength, F_e, is used for both parallel and perpendicular to grain loading.

11.3.2.2 Dowel bearing strengths, F_e, for wood structural panels are provided in Table 11.3.2B.

11.3.2.3 Dowel bearing strengths, F_e, for structural composite lumber shall be obtained from the manufacturer's literature or code evaluation report.

11.3.2.4 When dowel-type fasteners with $D \geq 1/4"$ are inserted into the end grain of the main member, with the fastener axis parallel to the wood fibers, $F_{e\perp}$ shall be used in determination of the dowel bearing strength of the main member, F_{em}.

11.3.3 Dowel Bearing Strength at an Angle to Grain

When a member in a connection is loaded at an angle to grain, the dowel bearing strength, $F_{e\theta}$, for the member shall be determined as follows (see Appendix J):

$$F_{e\theta} = \frac{F_{e\parallel} F_{e\perp}}{F_{e\parallel} \sin^2 \theta + F_{e\perp} \cos^2 \theta} \qquad (11.3\text{-}11)$$

where:

θ = angle between direction of load and direction of grain (longitudinal axis of member).

Figure 11B Single Shear Bolted Connections

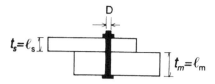

Figure 11C Double Shear Bolted Connections

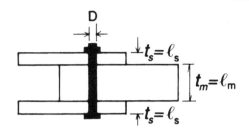

11.3.4 Dowel Bearing Length

11.3.4.1 Dowel bearing length in the side member(s) and main member, ℓ_s and ℓ_m, represent the length of dowel bearing perpendicular to the application of load. The length of dowel bearing shall not include the tapered tip of a fastener for fastener penetration lengths less than 10D.

11.3.5 Dowel Bending Yield Strength

11.3.5.1 Reference design values for bolts, lag screws, wood screws, nails, and spikes are based on bending yield strengths provided in Tables 11A through 11R.

11.3.5.2 Dowel bending yield strengths, F_{yb}, used in calculation of reference design values shall be based on yield strength derived using methods provided in ASTM F 1575 or the tensile yield strength derived using procedures of ASTM F 606.

11.3.6 Dowel Diameter

11.3.6.1 When used in Tables 11.3-1A and 11.3-1B, the fastener diameter shall be taken as D for unthreaded full-body diameter fasteners and D_r for reduced body diameter fasteners or threaded fasteners except as provided in 11.3.6.2. For bolts meeting the requirements of ANSI/ASME Standard B18.2.1 for full-body diameter bolts, the fastener diameter shall be taken as D (see Appendix L).

Table 11.3.2 Dowel Bearing Strengths

Specific[1] Gravity, G	F_e D<1/4"	$F_{e\parallel}$ D≥1/4"	$F_{e\perp}$ D=1/4"	D=5/16"	D=3/8"	D=7/16"	D=1/2"	D=5/8"	D=3/4"	D=7/8"	D=1"
0.73	9300	8200	7750	6900	6300	5850	5450	4900	4450	4150	3850
0.72	9050	8050	7600	6800	6200	5750	5350	4800	4350	4050	3800
0.71	8850	7950	7400	6650	6050	5600	5250	4700	4300	3950	3700
0.70	8600	7850	7250	6500	5950	5500	5150	4600	4200	3900	3650
0.69	8400	7750	7100	6350	5800	5400	5050	4500	4100	3800	3550
0.68	8150	7600	6950	6250	5700	5250	4950	4400	4050	3750	3500
0.67	7950	7500	6850	6100	5550	5150	4850	4300	3950	3650	3400
0.66	7750	7400	6700	5950	5450	5050	4700	4200	3850	3550	3350
0.65	7500	7300	6550	5850	5350	4950	4600	4150	3750	3500	3250
0.64	7300	7150	6400	5700	5200	4850	4500	4050	3700	3400	3200
0.63	7100	7050	6250	5600	5100	4700	4400	3950	3600	3350	3100
0.62	6900	6950	6100	5450	5000	4600	4300	3850	3500	3250	3050
0.61	6700	6850	5950	5350	4850	4500	4200	3750	3450	3200	3000
0.60	6500	6700	5800	5200	4750	4400	4100	3700	3350	3100	2900
0.59	6300	6600	5700	5100	4650	4300	4000	3600	3300	3050	2850
0.58	6100	6500	5550	4950	4500	4200	3900	3500	3200	2950	2750
0.57	5900	6400	5400	4850	4400	4100	3800	3400	3100	2900	2700
0.56	5700	6250	5250	4700	4300	4000	3700	3350	3050	2800	2650
0.55	5550	6150	5150	4600	4200	3900	3650	3250	2950	2750	2550
0.54	5350	6050	5000	4450	4100	3750	3550	3150	2900	2650	2500
0.53	5150	5950	4850	4350	3950	3650	3450	3050	2800	2600	2450
0.52	5000	5800	4750	4250	3850	3550	3350	3000	2750	2550	2350
0.51	4800	5700	4600	4100	3750	3450	3250	2900	2650	2450	2300
0.50	4650	5600	4450	4000	3650	3400	3150	2800	2600	2400	2250
0.49	4450	5500	4350	3900	3550	3300	3050	2750	2500	2300	2150
0.48	4300	5400	4200	3750	3450	3200	3000	2650	2450	2250	2100
0.47	4150	5250	4100	3650	3350	3100	2900	2600	2350	2200	2050
0.46	4000	5150	3950	3550	3250	3000	2800	2500	2300	2100	2000
0.45	3800	5050	3850	3450	3150	2900	2700	2400	2200	2050	1900
0.44	3650	4950	3700	3300	3050	2800	2600	2350	2150	2000	1850
0.43	3500	4800	3600	3200	2950	2700	2550	2250	2050	1900	1800
0.42	3350	4700	3450	3100	2850	2600	2450	2200	2000	1850	1750
0.41	3200	4600	3350	3000	2750	2550	2350	2100	1950	1800	1650
0.40	3100	4500	3250	2900	2650	2450	2300	2050	1850	1750	1600
0.39	2950	4350	3100	2800	2550	2350	2200	1950	1800	1650	1550
0.38	2800	4250	3000	2700	2450	2250	2100	1900	1750	1600	1500
0.37	2650	4150	2900	2600	2350	2200	2050	1850	1650	1550	1450
0.36	2550	4050	2750	2500	2250	2100	1950	1750	1600	1500	1400
0.35	2400	3900	2650	2400	2150	2000	1900	1700	1550	1400	1350
0.34	2300	3800	2550	2300	2100	1950	1800	1600	1450	1350	1300
0.33	2150	3700	2450	2200	2000	1850	1750	1550	1400	1300	1200
0.32	2050	3600	2350	2100	1900	1750	1650	1500	1350	1250	1150
0.31	1900	3450	2250	2000	1800	1700	1600	1400	1300	1200	1100

1. Specific gravity based on weight and volume when oven-dry (see Table 11.3.2A). Different specific gravities (G) are possible for different grades of MSR and MEL lumber (see Table 4C, Footnote 2).

2. $F_{e\parallel} = 11200G$; $F_{e\perp} = 6100G^{1.45}/\sqrt{D}$; F_e for D < 1/4" = 16600 $G^{1.84}$; Tabulated values are rounded to the nearest 50 psi.

DOWEL-TYPE FASTENERS

11

Table 11.3.2A   Assigned Specific Gravities

Species Combination	Specific[1] Gravity, G	Species Combinations of MSR and MEL Lumber	Specific[1] Gravity, G
Aspen	0.39	Douglas Fir-Larch	
Alaska Cedar	0.47	E=1,900,000 psi and lower grades of MSR	0.50
Alaska Hemlock	0.46	E=2,000,000 psi grades of MSR	0.51
Alaska Spruce	0.41	E=2,100,000 psi grades of MSR	0.52
Alaska Yellow Cedar	0.46	E=2,200,000 psi grades of MSR	0.53
Balsam Fir	0.36	E=2,300,000 psi grades of MSR	0.54
Beech-Birch-Hickory	0.71	E=2,400,000 psi grades of MSR	0.55
Coast Sitka Spruce	0.39	Douglas Fir-Larch (North)	
Cottonwood	0.41	E=1,900,000 psi and lower grades of MSR and MEL	0.49
Douglas Fir-Larch	0.50	E=2,000,000 psi to 2,200,000 psi grades of MSR and MEL	0.53
Douglas Fir-Larch (North)	0.49	E=2,300,000 psi and higher grades of MSR and MEL	0.57
Douglas Fir-South	0.46	Douglas Fir-Larch (South)	
Eastern Hemlock	0.41	E=1,000,000 psi and higher grades of MSR	0.46
Eastern Hemlock-Balsam Fir	0.36	Engelmann Spruce-Lodgepole Pine	
Eastern Hemlock-Tamarack	0.41	E=1,400,000 psi and lower grades of MSR	0.38
Eastern Hemlock-Tamarack (North)	0.47	E=1,500,000 psi and higher grades of MSR	0.46
Eastern Softwoods	0.36	Hem-Fir	
Eastern Spruce	0.41	E=1,500,000 psi and lower grades of MSR	0.43
Eastern White Pine	0.36	E=1,600,000 psi grades of MSR	0.44
Engelmann Spruce-Lodgepole Pine	0.38	E=1,700,000 psi grades of MSR	0.45
Hem-Fir	0.43	E=1,800,000 psi grades of MSR	0.46
Hem-Fir (North)	0.46	E=1,900,000 psi grades of MSR	0.47
Mixed Maple	0.55	E=2,000,000 psi grades of MSR	0.48
Mixed Oak	0.68	E=2,100,000 psi grades of MSR	0.49
Mixed Southern Pine	0.51	E=2,200,000 psi grades of MSR	0.50
Mountain Hemlock	0.47	E=2,300,000 psi grades of MSR	0.51
Northern Pine	0.42	E=2,400,000 psi grades of MSR	0.52
Northern Red Oak	0.68	Hem-Fir (North)	
Northern Species	0.35	E=1,000,000 psi and higher grades of MSR and MEL	0.46
Northern White Cedar	0.31	Southern Pine	
Ponderosa Pine	0.43	E=1,700,000 psi and lower grades of MSR and MEL	0.55
Red Maple	0.58	E=1,800,000 psi and higher grades of MSR and MEL	0.57
Red Oak	0.67	Spruce-Pine-Fir	
Red Pine	0.44	E=1,700,000 psi and lower grades of MSR and MEL	0.42
Redwood, close grain	0.44	E=1,800,000 psi and 1,900,000 grades of MSR and MEL	0.46
Redwood, open grain	0.37	E=2,000,000 psi and higher grades of MSR and MEL	0.50
Sitka Spruce	0.43	Spruce-Pine-Fir (South)	
Southern Pine	0.55	E=1,100,000 psi and lower grades of MSR	0.36
Spruce-Pine-Fir	0.42	E=1,200,000 psi to 1,900,000 psi grades of MSR	0.42
Spruce-Pine-Fir (South)	0.36	E=2,000,000 psi and higher grades of MSR	0.50
Western Cedars	0.36	Western Cedars	
Western Cedars (North)	0.35	E=1,000,000 psi and higher grades of MSR	0.36
Western Hemlock	0.47	Western Woods	
Western Hemlock (North)	0.46	E=1,000,000 psi and higher grades of MSR	0.36
Western White Pine	0.40		
Western Woods	0.36		
White Oak	0.73		
Yellow Poplar	0.43		

1. Specific gravity based on weight and volume when oven-dry. Different specific gravities (G) are possible for different grades of MSR and MEL lumber (see Table 4C, Footnote 2).

Table 11.3.2B Dowel Bearing Strengths for Wood Structural Panels

Wood Structural Panel	Specific[1] Gravity, G	Dowel Bearing Strength, F_e, in pounds per square inch (psi)
Plywood		
Structural 1, Marine	0.50	4650
Other Grades[1]	0.42	3350
Oriented Strand Board		
All Grades	0.50	4650

1. Use G = 0.42 when species of the plies is not known. When species of the plies is known, specific gravity listed for the actual species and the corresponding dowel bearing strength may be used, or the weighted average may be used for mixed species.

11.3.6.2 For threaded full-body fasteners (see Appendix L), D shall be permitted to be used in lieu of D_r when the bearing length of the threads does not exceed ¼ of the full bearing length in the member holding the threads. Alternatively, a more detailed analysis accounting for the moment and bearing resistance of the threaded portion of the fastener shall be permitted (see Appendix I).

11.3.7 Asymmetric Three Member Connections, Double Shear

Reference design values, Z, for asymmetric three member connections shall be the minimum computed yield mode value for symmetric double shear connections using the smaller dowel bearing length in the side member as ℓ_s and the minimum dowel diameter, D, occurring in either of the connection shear planes.

Figure 11D Multiple Shear Bolted Connections

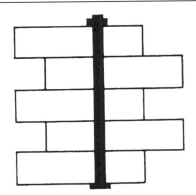

11.3.8 Multiple Shear Connections

For a connection with four or more members (see Figure 11D), each shear plane shall be evaluated as a single shear connection. The reference design value for the connection shall be the lowest reference design value for any single shear plane, multiplied by the number of shear planes.

11.3.9 Load at an Angle to Fastener Axis

11.3.9.1 When the applied load in a single shear (two member) connection is at an angle (other than 90°) with the fastener axis, the fastener lengths in the two members shall be designated ℓ_s and ℓ_m (see Figure 11E). The component of the load acting at 90° with the fastener axis shall not exceed the adjusted design value, Z', for a connection in which two members at 90° with the fastener axis have thicknesses $t_s = \ell_s$ and $t_m = \ell_m$. Ample bearing area shall be provided to resist the load component acting parallel to the fastener axis.

11.3.9.2 For toe-nailed connections, use the minimum of t_s or L/3 for ℓ_s (see Figure 11A).

11.3.10 Drift Bolts and Drift Pins

Adjusted lateral design values for drift bolts and drift pins driven in the side grain of wood shall not exceed 75% of the adjusted lateral design values for common bolts of the same diameter and length in main member.

Figure 11E Shear Area for Bolted Connections

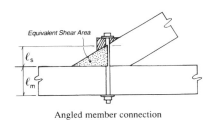

Angled member connection

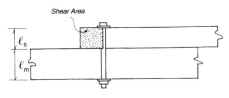

Parallel member connection

DOWEL-TYPE FASTENERS

11

11.4 Combined Lateral and Withdrawal Loads

11.4.1 Lag Screws and Wood Screws

When a lag screw or wood screw is subjected to combined lateral and withdrawal loading, as when the fastener is inserted perpendicular to the fiber and the load acts at an angle, α, to the wood surface (see Figure 11F), the adjusted design value shall be determined as follows (see Appendix J):

$$Z_\alpha' = \frac{(W'p)Z'}{(W'p)\cos^2\alpha + Z'\sin^2\alpha} \qquad (11.4\text{-}1)$$

where:

α = angle between wood surface and direction of applied load

p = length of thread penetration in main member, in.

11.4.2 Nails and Spikes

When a nail or spike is subjected to combined lateral and withdrawal loading, as when the nail or spike is inserted perpendicular to the fiber and the load acts at an angle, α, to the wood surface, the adjusted design value shall be determined as follows:

$$Z_\alpha' = \frac{(W'p)Z'}{(W'p)\cos\alpha + Z'\sin\alpha} \qquad (11.4\text{-}2)$$

where:

α = angle between wood surface and direction of applied load

p = length of penetration in main member, in.

Figure 11F   Combined Lateral and Withdrawal Loading

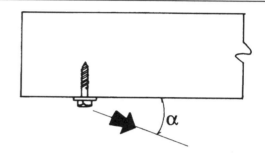

11.5 Adjustment of Reference Design Values

11.5.1 Geometry Factor, C_Δ

11.5.1.1 When D < 1/4", C_Δ = 1.0.

11.5.1.2 When D ≥ 1/4" and the end distance or spacing provided for dowel-type fasteners is less than the minimum required for C_Δ = 1.0 for any condition in (a), (b), or (c), reference design values shall be multiplied by the smallest applicable geometry factor, C_Δ, determined in (a), (b), or (c). The smallest geometry factor for any fastener in a group shall apply to all fasteners in the group. For multiple shear connections or for asymmetric three member connections, the smallest geometry factor, C_Δ, for any shear plane shall apply to all fasteners in the connection. Provisions for C_Δ are based on an assumption that edge distance and spacing between rows of fasteners is in accordance with Table 11.5.1A and Table 11.5.1D and applicable requirements of 11.1.

Table 11.5.1A   Edge Distance Requirements [1,2]

Direction of Loading	Minimum Edge Distance
Parallel to Grain:	
when $\ell/D \le 6$	1.5D
when $\ell/D > 6$	1.5D or ½ the spacing between rows, whichever is greater
Perpendicular to Grain:[2]	
loaded edge	4D
unloaded edge	1.5D

1. The ℓ/D ratio used to determine the minimum edge distance shall be the lesser of:
       (a) length of fastener in wood main member/D = ℓ_m/D
       (b) total length of fastener in wood side member(s)/D = ℓ_s/D
2. Heavy or medium concentrated loads shall not be suspended below the neutral axis of a single sawn lumber or structural glued laminated timber beam except where mechanical or equivalent reinforcement is provided to resist tension stresses perpendicular to grain (see 3.8.2 and 10.1.3).

Figure 11G Bolted Connection Geometry

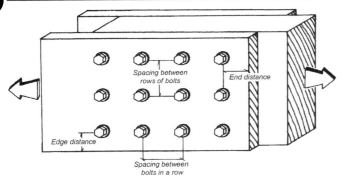

Parallel to grain loading in all wood members ($Z_{\parallel}$)

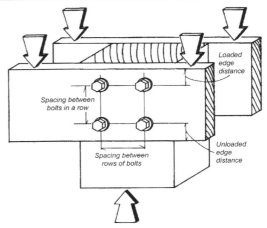

Perpendicular to grain loading in the side member
and parallel to grain loading in the main member ($Z_{s\perp}$)

(a) When dowel-type fasteners are used and the actual end distance for parallel or perpendicular to grain loading is greater than or equal to the minimum end distance (see Table 11.5.1B) for $C_\Delta = 0.5$, but less than the minimum end distance for $C_\Delta = 1.0$, the geometry factor, C_Δ, shall be determined as follows:

$$C_\Delta = \frac{\text{actual end distance}}{\text{minimum end distance for } C_\Delta = 1.0}$$

(b) For loading at an angle to the fastener, when dowel-type fasteners are used, the minimum shear area for $C_\Delta = 1.0$ shall be equivalent to the shear area for a parallel member connection with minimum end distance for $C_\Delta = 1.0$ (see Table 11.5.1B and Figure 11E). The minimum shear area for $C_\Delta = 0.5$ shall be equivalent to ½ the minimum shear area for $C_\Delta = 1.0$. When the actual shear area is greater than or equal to the minimum shear area for $C_\Delta = 0.5$, but less than the minimum shear area for $C_\Delta = 1.0$, the geometry factor, C_Δ, shall be determined as follows:

$$C_\Delta = \frac{\text{actual shear area}}{\text{minimum shear area for } C_\Delta = 1.0}$$

(c) When the actual spacing between dowel-type fasteners in a row for parallel or perpendicular to grain loading is greater than or equal to the minimum spacing (see Table 11.5.1C), but less than the minimum spacing for $C_\Delta = 1.0$, the geometry factor, C_Δ, shall be determined as follows:

$$C_\Delta = \frac{\text{actual spacing}}{\text{minimum spacing for } C_\Delta = 1.0}$$

Table 11.5.1B End Distance Requirements

	End Distances	
Direction of Loading	Minimum end distance for $C_\Delta = 0.5$	Minimum end distance for $C_\Delta = 1.0$
Perpendicular to Grain	2D	4D
Parallel to Grain, Compression: (fastener bearing away from member end)	2D	4D
Parallel to Grain, Tension: (fastener bearing toward member end)		
for softwoods	3.5D	7D
for hardwoods	2.5D	5D

Table 11.5.1C Spacing Requirements for Fasteners in a Row

	Spacing	
Direction of Loading	Minimum spacing	Minimum spacing for $C_\Delta = 1.0$
Parallel to Grain	3D	4D
Perpendicular to Grain	3D	Required spacing for attached members

DOWEL-TYPE FASTENERS

11

Table 11.5.1D Spacing Requirements Between Rows [1,2]

Direction of Loading	Minimum Spacing
Parallel to Grain	1.5D
Perpendicular to Grain:	
when $\ell/D \leq 2$	2.5D
when $2 < \ell/D < 6$	$(5\ell + 10D) / 8$
when $\ell/D \geq 6$	5D

1. The ℓ/D ratio used to determine the minimum edge distance shall be the lesser of:
   (a) length of fastener in wood main member/D = ℓ_m/D
   (b) total length of fastener in wood side member(s)/D = ℓ_s/D
2. The spacing between outer rows of fasteners paralleling the member on a single splice plate shall not exceed 5" (see Figure 11H).

11.5.2 End Grain Factor, C_{eg}

11.5.2.1 When lag screws are loaded in withdrawal from end grain, the reference withdrawal design values, W, shall be multiplied by the end grain factor, C_{eg} = 0.75.

11.5.2.2 When dowel-type fasteners are inserted in the end grain of the main member, with the fastener axis parallel to the wood fibers, reference lateral design values, Z, shall be multiplied by the end grain factor, C_{eg} = 0.67.

11.5.3 Diaphragm Factor, C_{di}

When nails or spikes are used in diaphragm construction, reference lateral design values, Z, shall be multiplied by the diaphragm factor, C_{di} = 1.1.

11.5.4 Toe-Nail Factor, C_{tn}

11.5.4.1 When toe-nailed connections are used, reference withdrawal design values, W, for the nails or spikes shall be multiplied by the toe-nail factor, C_{tn} = 0.67. The wet service factor, C_M, shall not apply for toe-nailed connections loaded in withdrawal.

11.5.4.2 When toe-nailed connections are used, reference lateral design values, Z, shall be multiplied by the toe-nail factor, C_{tn} = 0.83.

Table 11.5.1E Edge and End Distance and Spacing Requirements for Lag Screws Loaded in Withdrawal and Not Loaded Laterally

Orientation	Minimum Distance/Spacing
Edge Distance	1.5D
End Distance	4D
Spacing	4D

Figure 11H Spacing Between Outer Rows of Bolts

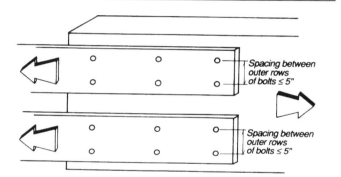

Spacing between outer rows of bolts ≤ 5"

Spacing between outer rows of bolts ≤ 5"

11.6 Multiple Fasteners

11.6.1 Symmetrically Staggered Fasteners

When a connection contains multiple fasteners, fasteners shall be staggered symmetrically in members loaded perpendicular to grain whenever possible (see 3.1.2.2 and 10.3.6.2 for special design provisions when bolts, lag screws, or drift pins are staggered).

11.6.2 Fasteners Loaded at an Angle to Grain

When a multiple fastener connection is loaded at an angle to grain, the gravity axis of each member shall pass through the center of resistance of the group of fasteners to insure uniform stress in the main member and a uniform distribution of load to all fasteners.

11.6.3 Local Stresses in Connections

Local stresses in connections using multiple fasteners shall be evaluated in accordance with principles of engineering mechanics (see 10.1.2).

DOWEL-TYPE FASTENERS

11

Table 11A BOLTS: Reference Lateral Design Values (Z) for Single Shear (two member) Connections[1,2]

for sawn lumber or SCL with both members of identical specific gravity

t_m (in.)	t_s (in.)	D (in.)	G=0.67 Red Oak $Z_{\parallel}$	$Z_{s\perp}$	$Z_{m\perp}$	$Z_{\perp}$	G=0.55 Mixed Maple / Southern Pine $Z_{\parallel}$	$Z_{s\perp}$	$Z_{m\perp}$	$Z_{\perp}$	G=0.50 Douglas Fir-Larch $Z_{\parallel}$	$Z_{s\perp}$	$Z_{m\perp}$	$Z_{\perp}$	G=0.49 Douglas Fir-Larch(N) $Z_{\parallel}$	$Z_{s\perp}$	$Z_{m\perp}$	$Z_{\perp}$	G=0.46 Douglas Fir(S) / Hem-Fir(N) $Z_{\parallel}$	$Z_{s\perp}$	$Z_{m\perp}$	$Z_{\perp}$
1-1/2	1-1/2	1/2	650	420	420	330	530	330	330	250	480	300	300	220	470	290	290	210	440	270	270	190
		5/8	810	500	500	370	660	400	400	280	600	360	360	240	590	350	350	240	560	320	320	220
		3/4	970	580	580	410	800	460	460	310	720	420	420	270	710	400	400	260	670	380	380	240
		7/8	1130	660	660	440	930	520	520	330	850	470	470	290	830	460	460	280	780	420	420	250
		1	1290	740	740	470	1060	580	580	350	970	530	530	310	950	510	510	300	890	480	480	280
1-3/4	1-3/4	1/2	760	490	490	390	620	390	390	290	560	350	350	250	550	340	340	250	520	320	320	230
		5/8	940	590	590	430	770	470	470	330	700	420	420	280	690	410	410	280	650	380	380	250
		3/4	1130	680	680	480	930	540	540	360	850	480	480	310	830	470	470	300	780	440	440	280
		7/8	1320	770	770	510	1080	610	610	390	970	550	550	340	970	530	530	340	910	500	500	300
		1	1510	860	860	550	1240	680	680	410	1130	610	610	360	1110	600	600	350	1040	560	560	320
2-1/2	1-1/2	1/2	770	480	540	440	660	400	420	350	610	370	370	310	610	360	360	300	580	340	330	270
		5/8	1070	660	630	520	930	560	490	390	850	520	430	340	830	520	420	330	780	470	390	300
		3/4	1360	890	720	570	1120	660	560	430	1020	590	500	380	1000	560	480	360	940	520	450	330
		7/8	1590	960	800	620	1300	720	620	470	1190	630	550	410	1170	600	540	390	1090	550	500	360
		1	1820	1020	870	660	1490	770	680	490	1360	680	610	440	1330	650	590	420	1250	600	550	390
3-1/2	1-1/2	1/2	770	480	560	440	660	400	470	360	610	370	430	330	610	360	420	320	580	340	400	310
		5/8	1070	660	760	590	940	560	620	500	880	520	540	460	870	520	530	450	830	470	490	410
		3/4	1450	890	900	770	1270	660	690	580	1200	590	610	510	1190	560	590	490	1140	520	550	450
		7/8	1890	960	990	830	1680	720	770	630	1590	630	680	550	1570	600	650	530	1470	550	600	480
		1	2410	1020	1080	890	2010	770	830	670	1830	680	740	590	1790	650	710	560	1680	600	660	520
	1-3/4	1/2	830	510	590	480	720	420	510	390	670	380	470	350	660	380	460	340	620	360	440	320
		5/8	1160	680	820	620	1000	580	640	520	930	530	560	460	920	530	550	450	880	500	510	410
		3/4	1530	900	940	780	1330	770	720	580	1250	680	640	520	1240	660	620	500	1190	600	580	460
		7/8	1970	1120	1040	840	1730	840	810	640	1620	740	710	550	1590	700	690	530	1490	640	640	490
		1	2480	1190	1130	900	2030	890	880	670	1850	790	780	590	1820	750	760	570	1700	700	700	530
	3-1/2	1/2	830	590	590	530	750	520	520	460	720	490	490	430	710	480	480	420	690	460	460	410
		5/8	1290	880	880	780	1170	780	780	650	1120	700	700	560	1110	690	690	550	1070	650	650	500
		3/4	1860	1190	1190	950	1690	960	960	710	1610	870	870	630	1600	850	850	600	1540	800	800	560
		7/8	2540	1410	1410	1030	2170	1160	1160	780	1970	1060	1060	680	1940	1040	1040	650	1810	980	980	590
		1	3020	1670	1670	1100	2480	1360	1360	820	2260	1230	1230	720	2210	1190	1190	690	2070	1110	1110	640
5-1/4	1-1/2	5/8	1070	660	760	590	940	560	640	500	880	520	590	460	870	520	590	450	830	470	560	430
		3/4	1450	890	990	780	1270	660	850	660	1200	590	790	590	1190	560	780	560	1140	520	740	520
		7/8	1890	960	1260	960	1680	720	1060	720	1590	630	940	630	1570	600	900	600	1520	550	830	550
		1	2410	1020	1500	1020	2150	770	1140	770	2050	680	1010	680	2030	650	970	650	1930	600	910	600
	1-3/4	5/8	1160	680	820	620	1000	580	690	520	930	530	630	470	920	530	630	470	880	500	590	440
		3/4	1530	900	1050	800	1330	770	800	680	1250	680	830	630	1240	660	810	620	1190	600	780	590
		7/8	1970	1120	1320	1020	1730	840	1090	840	1640	740	960	740	1620	700	920	700	1550	640	850	640
		1	2480	1190	1530	1190	2200	890	1170	890	2080	790	1040	790	2060	750	1000	750	1990	700	930	700
	3-1/2	5/8	1290	880	880	780	1170	780	780	680	1120	700	730	630	1110	690	720	620	1070	650	690	580
		3/4	1860	1190	1240	1080	1690	960	1090	850	1610	870	1030	780	1600	850	1010	750	1540	800	970	710
		7/8	2540	1410	1640	1260	2300	1160	1380	1000	2190	1060	1230	870	2170	1040	1190	840	2060	980	1100	770
		1	3310	1670	1940	1420	2870	1390	1520	1060	2660	1290	1360	940	2630	1260	1320	900	2500	1210	1230	830
5-1/2	1-1/2	5/8	1070	660	760	590	940	560	640	500	880	520	590	460	870	520	590	450	830	470	560	430
		3/4	1450	890	990	780	1270	660	850	660	1200	590	790	590	1190	560	780	560	1140	520	740	520
		7/8	1890	960	1260	960	1680	720	1090	720	1590	630	980	630	1570	600	940	600	1520	550	860	550
		1	2410	1020	1560	1020	2150	770	1190	770	2050	680	1060	680	2030	650	1010	650	1930	600	940	600
	3-1/2	5/8	1290	880	880	780	1170	780	780	680	1120	700	730	630	1110	690	720	620	1070	650	690	580
		3/4	1860	1190	1240	1080	1690	960	1090	850	1610	870	1030	780	1600	850	1010	750	1540	800	970	710
		7/8	2540	1410	1640	1260	2300	1160	1410	1020	2190	1060	1260	910	2170	1040	1220	870	2060	980	1130	790
		1	3310	1670	1980	1470	2870	1390	1550	1100	2660	1290	1390	970	2630	1260	1340	930	2500	1210	1250	860
7-1/2	1-1/2	5/8	1070	660	760	590	940	560	640	500	880	520	590	460	870	520	590	450	830	470	560	430
		3/4	1450	890	990	780	1270	660	850	660	1200	590	790	590	1190	560	780	560	1140	520	740	520
		7/8	1890	960	1260	960	1680	720	1090	720	1590	630	1010	630	1570	600	990	600	1520	550	950	550
		1	2410	1020	1560	1020	2150	770	1350	770	2050	680	1270	680	2030	650	1240	650	1930	600	1190	600
	3-1/2	5/8	1290	880	880	780	1170	780	780	680	1120	700	730	630	1110	690	720	620	1070	650	690	580
		3/4	1860	1190	1240	1080	1690	960	1090	850	1610	870	1030	780	1600	850	1010	750	1540	800	970	710
		7/8	2540	1410	1640	1260	2300	1160	1450	1020	2190	1060	1360	930	2170	1040	1340	900	2060	980	1280	850
		1	3310	1670	2090	1470	2870	1390	1830	1210	2660	1290	1630	1110	2630	1260	1570	1080	2500	1210	1470	1030

1. Tabulated lateral design values (Z) for bolted connections shall be multiplied by all applicable adjustment factors (see Table 10.3.1).
2. Tabulated lateral design values (Z) are for "full diameter" bolts (see Appendix L) with bending yield strength (F_{yb}) of 45,000 psi.

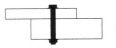

Table 11A (Cont.) BOLTS: Reference Lateral Design Values (Z) for Single Shear (two member) Connections[1,2]

for sawn lumber or SCL with both members of identical specific gravity

Thickness Main Member t_m in.	Thickness Side Member t_s in.	Bolt Diameter D in.	G=0.43 Hem-Fir $Z_\parallel$ lbs.	$Z_{s\perp}$ lbs.	$Z_{m\perp}$ lbs.	$Z_\perp$ lbs.	G=0.42 Spruce-Pine-Fir $Z_\parallel$ lbs.	$Z_{s\perp}$ lbs.	$Z_{m\perp}$ lbs.	$Z_\perp$ lbs.	G=0.37 Redwood (open grain) $Z_\parallel$ lbs.	$Z_{s\perp}$ lbs.	$Z_{m\perp}$ lbs.	$Z_\perp$ lbs.	G=0.36 Eastern Softwoods, Spruce-Pine-Fir(S), Western Cedars, Western Woods $Z_\parallel$ lbs.	$Z_{s\perp}$ lbs.	$Z_{m\perp}$ lbs.	$Z_\perp$ lbs.	G=0.35 Northern Species $Z_\parallel$ lbs.	$Z_{s\perp}$ lbs.	$Z_{m\perp}$ lbs.	$Z_\perp$ lbs.
1-1/2	1-1/2	1/2	410	250	250	180	410	240	240	170	360	210	210	140	350	200	200	130	340	200	200	130
		5/8	520	300	300	190	510	290	290	190	450	250	250	160	440	240	240	150	420	240	240	150
		3/4	620	350	350	210	610	340	340	210	540	290	290	170	520	280	280	170	500	270	270	160
		7/8	720	390	390	230	710	380	380	220	630	330	330	190	610	320	320	180	590	310	310	170
		1	830	440	440	250	810	430	430	240	720	370	370	200	700	360	360	190	670	350	350	190
1-3/4	1-3/4	1/2	480	290	290	210	470	280	280	200	420	250	250	170	410	240	240	160	390	230	230	150
		5/8	600	350	350	230	590	340	340	220	520	290	290	190	510	280	280	180	490	270	270	170
		3/4	720	400	400	250	710	390	390	240	630	340	340	200	610	330	330	190	590	320	320	190
		7/8	850	460	460	270	830	450	450	260	730	390	390	220	710	380	380	210	690	360	360	200
		1	970	510	510	290	950	500	500	280	840	430	430	230	820	420	420	230	790	410	410	220
2-1/2	1-1/2	1/2	550	320	310	250	540	320	300	240	500	290	250	200	490	280	240	190	470	280	240	180
		5/8	730	420	360	270	710	410	350	270	630	350	300	220	610	330	290	210	590	320	280	210
		3/4	870	460	410	300	850	450	400	290	750	370	340	240	740	360	330	230	710	350	320	230
		7/8	1020	500	450	320	1000	490	440	310	880	410	380	260	860	390	370	250	830	370	350	240
		1	1160	540	500	350	1140	530	490	340	1010	440	420	280	980	420	410	270	940	410	390	260
3-1/2	1-1/2	1/2	550	320	380	290	540	320	370	280	500	290	320	250	490	280	300	250	480	280	290	240
		5/8	790	420	440	370	780	410	430	360	720	350	370	300	710	330	350	290	700	320	340	280
		3/4	1100	460	500	400	1080	450	480	390	1010	370	410	320	990	360	400	310	950	350	380	300
		7/8	1370	500	550	430	1340	490	540	420	1180	410	460	350	1160	390	440	340	1110	370	420	320
		1	1570	540	600	470	1530	530	590	460	1350	440	500	380	1320	420	480	370	1270	410	470	350
	1-3/4	1/2	590	340	400	300	580	330	390	290	530	300	330	260	520	290	320	250	510	280	310	250
		5/8	840	480	460	370	820	470	450	360	760	400	390	310	740	380	370	290	730	370	360	280
		3/4	1130	540	520	410	1120	530	510	400	1030	430	430	330	1000	420	420	320	970	410	410	310
		7/8	1390	580	580	440	1360	570	570	430	1200	470	480	360	1170	460	470	350	1130	430	440	320
		1	1590	630	640	480	1550	610	630	460	1370	510	530	380	1340	490	520	370	1290	470	500	360
	3-1/2	1/2	660	440	440	390	660	430	430	380	620	400	400	330	610	390	390	310	600	380	380	310
		5/8	1040	600	600	450	1020	590	590	440	960	520	520	370	950	500	500	350	930	490	490	340
		3/4	1450	740	740	500	1420	730	730	480	1250	650	650	400	1220	630	630	390	1180	620	620	370
		7/8	1690	910	910	540	1660	890	890	520	1460	770	770	440	1430	750	750	420	1370	720	720	390
		1	1930	1030	1030	580	1890	1000	1000	560	1670	870	870	470	1630	840	840	450	1570	810	810	430
5-1/4	1-1/2	5/8	790	420	530	410	780	410	520	400	720	350	470	350	710	330	460	330	700	320	450	320
		3/4	1100	460	690	460	1080	450	670	450	1010	370	560	370	990	360	540	360	970	350	530	350
		7/8	1460	500	750	500	1440	490	730	490	1350	410	620	410	1330	390	600	390	1280	370	560	370
		1	1800	540	820	540	1760	530	800	530	1560	440	670	440	1520	420	650	420	1460	410	630	410
	1-3/4	5/8	840	480	560	410	820	470	550	410	760	400	500	370	740	380	480	360	730	370	470	350
		3/4	1130	540	700	540	1120	530	680	530	1040	430	570	430	1020	420	560	420	1000	410	540	410
		7/8	1490	580	770	580	1470	570	750	570	1370	470	640	470	1350	460	620	460	1320	430	580	430
		1	1910	630	850	630	1890	610	820	610	1760	510	690	510	1740	490	670	490	1700	470	650	470
	3-1/2	5/8	1040	600	660	530	1020	590	650	520	960	520	610	460	950	500	590	440	930	490	580	430
		3/4	1490	740	900	640	1480	730	880	620	1390	650	750	520	1370	630	730	500	1330	620	710	480
		7/8	1950	920	1010	690	1920	910	990	670	1740	820	850	560	1710	800	830	550	1660	770	780	510
		1	2370	1140	1130	750	2330	1120	1100	730	2120	1020	940	600	2080	980	910	580	2030	950	880	560
5-1/2	1-1/2	5/8	790	420	530	410	780	410	520	400	720	350	470	350	710	330	460	330	700	320	450	320
		3/4	1100	460	700	460	1080	450	690	450	1010	370	580	370	990	360	570	360	970	350	550	350
		7/8	1460	500	780	500	1440	490	760	490	1350	410	650	410	1330	390	630	390	1280	370	590	370
		1	1800	540	860	540	1760	530	830	530	1560	440	700	440	1520	420	680	420	1460	410	650	410
	3-1/2	5/8	1040	600	660	530	1020	590	650	520	960	520	610	460	950	500	590	440	930	490	580	430
		3/4	1490	740	920	650	1480	730	900	640	1390	650	770	530	1370	630	750	520	1330	620	720	500
		7/8	1950	920	1030	720	1920	910	1010	700	1740	820	870	590	1710	800	840	570	1660	770	800	530
		1	2370	1140	1150	780	2330	1120	1120	760	2120	1020	960	630	2080	980	930	600	2030	950	890	580
7-1/2	1-1/2	5/8	790	420	530	410	780	410	520	400	720	350	470	350	710	330	460	330	700	320	450	320
		3/4	1100	460	700	460	1080	450	690	450	1010	370	630	370	990	360	620	360	970	350	600	350
		7/8	1460	500	900	500	1440	490	890	490	1350	410	810	410	1330	390	800	390	1280	370	770	370
		1	1800	540	1130	540	1760	530	1110	530	1560	440	920	440	1520	420	890	420	1460	410	860	410
	3-1/2	5/8	1040	600	660	530	1020	590	650	520	960	520	610	460	950	500	590	440	930	490	580	430
		3/4	1490	740	920	650	1480	730	910	640	1390	650	840	560	1370	630	820	550	1330	620	810	540
		7/8	1950	920	1210	790	1920	910	1180	780	1740	820	1010	700	1710	800	980	680	1660	770	920	650
		1	2370	1140	1340	970	2330	1120	1300	950	2120	1020	1100	820	2080	980	1070	790	2030	950	1030	760

1. Tabulated lateral design values (Z) for bolted connections shall be multiplied by all applicable adjustment factors (see Table 10.3.1).
2. Tabulated lateral design values (Z) are for "full diameter" bolts (see Appendix L) with bending yield strength (F_{yb}) of 45,000 psi.

BOLTS

DOWEL-TYPE FASTENERS

11

Table 11B BOLTS: Reference Lateral Design Values (Z) for Single Shear (two member) Connections[1,2]

for sawn lumber or SCL with 1/4" ASTM A 36 steel side plate

Main Member t_m in.	Side Member t_s in.	Bolt Diameter D in.	G=0.67 Red Oak $Z_\parallel$ lbs.	$Z_\perp$ lbs.	G=0.55 Mixed Maple Southern Pine $Z_\parallel$ lbs.	$Z_\perp$ lbs.	G=0.50 Douglas Fir-Larch $Z_\parallel$ lbs.	$Z_\perp$ lbs.	G=0.49 Douglas Fir-Larch(N) $Z_\parallel$ lbs.	$Z_\perp$ lbs.	G=0.46 Douglas Fir(S) Hem-Fir(N) $Z_\parallel$ lbs.	$Z_\perp$ lbs.	G=0.43 Hem-Fir $Z_\parallel$ lbs.	$Z_\perp$ lbs.	G=0.42 Spruce-Pine-Fir $Z_\parallel$ lbs.	$Z_\perp$ lbs.	G=0.37 Redwood (open grain) $Z_\parallel$ lbs.	$Z_\perp$ lbs.	G=0.36 Eastern Softwoods Spruce-Pine-Fir(S) Western Cedars Western Woods $Z_\parallel$ lbs.	$Z_\perp$ lbs.	G=0.35 Northern Species $Z_\parallel$ lbs.	$Z_\perp$ lbs.
1-1/2	1/4	1/2	730	420	620	350	580	310	580	310	550	290	520	280	510	270	470	240	460	240	450	230
		5/8	910	480	780	400	730	360	720	360	690	340	650	320	640	320	590	290	580	280	560	270
		3/4	1090	550	940	450	870	420	860	410	820	390	780	360	770	360	710	320	690	320	680	310
		7/8	1270	600	1090	510	1020	470	1010	450	960	430	910	410	900	400	820	370	810	360	790	350
		1	1460	660	1250	550	1170	510	1150	500	1100	480	1040	450	1030	450	940	400	930	400	900	390
1-3/4	1/4	1/2	810	460	690	370	640	340	630	330	600	310	570	290	560	280	510	250	500	250	490	240
		5/8	1020	520	870	430	800	390	790	380	750	360	710	340	700	330	640	300	630	290	610	280
		3/4	1220	590	1040	480	960	440	950	430	900	410	860	380	840	370	770	330	750	330	730	320
		7/8	1420	650	1210	540	1130	490	1110	480	1050	450	1000	420	980	420	890	380	880	370	850	360
		1	1630	710	1380	580	1290	540	1270	520	1200	500	1140	470	1120	460	1020	410	1000	410	980	400
2-1/2	1/4	1/2	930	600	860	470	830	410	820	400	780	380	740	350	720	340	650	300	640	290	620	280
		5/8	1370	670	1150	530	1050	470	1040	470	980	430	920	400	910	390	810	340	800	330	770	320
		3/4	1640	750	1370	590	1270	530	1250	520	1180	490	1110	450	1090	440	980	380	960	370	930	360
		7/8	1910	820	1600	650	1480	590	1450	570	1370	530	1290	490	1270	480	1140	420	1120	410	1080	400
		1	2190	880	1830	700	1690	640	1660	620	1570	580	1480	540	1450	530	1300	460	1280	450	1240	440
3-1/2	1/4	1/2	930	620	860	550	830	510	820	510	800	480	770	450	770	430	720	370	720	360	710	350
		5/8	1370	860	1260	690	1210	610	1200	600	1160	550	1130	500	1120	490	1060	420	1050	410	1020	400
		3/4	1900	990	1740	760	1670	680	1660	660	1580	610	1480	560	1450	540	1290	460	1260	450	1220	440
		7/8	2530	1070	2170	840	1990	740	1950	710	1840	660	1720	610	1690	590	1510	510	1480	500	1430	470
		1	2980	1150	2480	890	2270	800	2230	770	2100	730	1970	660	1930	650	1720	560	1690	540	1630	530
5-1/4	1/4	5/8	1370	860	1260	760	1210	710	1200	700	1160	670	1130	640	1120	630	1060	580	1050	560	1030	540
		3/4	1900	1140	1740	1000	1670	940	1660	930	1610	860	1560	770	1550	760	1460	640	1450	620	1420	600
		7/8	2530	1460	2320	1190	2220	1050	2200	1010	2140	920	2070	840	2050	820	1940	700	1920	680	1890	640
		1	3260	1660	2980	1270	2860	1130	2840	1080	2750	1010	2670	920	2640	890	2490	750	2450	730	2360	710
5-1/2	1/4	5/8	1370	860	1260	760	1210	710	1200	700	1160	670	1130	640	1120	630	1060	580	1050	570	1030	560
		3/4	1900	1140	1740	1000	1670	940	1660	930	1610	890	1560	810	1550	790	1460	660	1450	640	1420	620
		7/8	2530	1460	2320	1240	2220	1090	2200	1050	2140	960	2070	880	2050	860	1940	730	1920	710	1890	660
		1	3260	1730	2980	1320	2860	1170	2840	1130	2750	1050	2670	950	2640	930	2490	780	2470	760	2420	740
7-1/2	1/4	5/8	1370	860	1260	760	1210	710	1200	700	1160	670	1130	640	1120	630	1060	580	1050	570	1030	560
		3/4	1900	1140	1740	1000	1670	940	1660	930	1610	890	1560	850	1550	840	1460	760	1450	750	1420	740
		7/8	2530	1460	2320	1280	2220	1210	2200	1180	2140	1130	2070	1080	2050	1070	1940	960	1920	930	1890	870
		1	3260	1820	2980	1590	2860	1500	2840	1470	2750	1400	2670	1270	2640	1230	2490	1030	2470	1000	2420	960
9-1/2	1/4	3/4	1900	1140	1740	1000	1670	940	1660	930	1610	890	1560	850	1550	840	1460	760	1450	750	1420	740
		7/8	2530	1460	2320	1280	2220	1210	2200	1180	2140	1130	2070	1080	2050	1070	1940	980	1920	970	1890	930
		1	3260	1820	2980	1590	2860	1500	2840	1470	2750	1420	2670	1350	2640	1330	2490	1220	2470	1200	2420	1180
11-1/2	1/4	7/8	2530	1460	2320	1280	2220	1210	2200	1180	2140	1130	2070	1080	2050	1070	1940	980	1920	970	1890	930
		1	3260	1820	2980	1590	2860	1500	2840	1470	2750	1420	2670	1350	2640	1330	2490	1220	2470	1200	2420	1180
13-1/2	1/4	1	3260	1820	2980	1590	2860	1500	2840	1470	2750	1420	2670	1350	2640	1330	2490	1220	2470	1200	2420	1180

1. Tabulated lateral design values (Z) for bolted connections shall be multiplied by all applicable adjustment factors (see Table 10.3.1).
2. Tabulated lateral design values (Z) are for "full diameter" bolts (see Appendix L) with bending yield strength (F_{yb}) of 45,000 psi and dowel bearing strength (F_e) of 87,000 psi for ASTM A 36 steel.

BOLTS

DOWEL-TYPE FASTENERS

11

Table 11C BOLTS: Reference Lateral Design Values (Z) for Single Shear (two member) Connections[1],[2]

for structural glued laminated timber main member with sawn lumber side member of identical specific gravity

Main Member t_m (in.)	Side Member t_s (in.)	Bolt Diameter D (in.)	G=0.55 Southern Pine $Z_{\parallel}$	$Z_{s\perp}$	$Z_{m\perp}$	$Z_{\perp}$	G=0.50 Douglas Fir-Larch $Z_{\parallel}$	$Z_{s\perp}$	$Z_{m\perp}$	$Z_{\perp}$	G=0.46 Douglas Fir(S) $Z_{\parallel}$	$Z_{s\perp}$	$Z_{m\perp}$	$Z_{\perp}$	G=0.43 Hem-Fir $Z_{\parallel}$	$Z_{s\perp}$	$Z_{m\perp}$	$Z_{\perp}$	G=0.42 Spruce-Pine-Fir $Z_{\parallel}$	$Z_{s\perp}$	$Z_{m\perp}$	$Z_{\perp}$	G=0.36 Spruce-Pine-Fir(S) Western Woods $Z_{\parallel}$	$Z_{s\perp}$	$Z_{m\perp}$	$Z_{\perp}$
2-1/2	1-1/2	1/2	-	-	-	360	610	370	370	310	580	340	330	270	550	320	310	250	540	320	300	240	490	280	240	190
		5/8	-	-	-	460	850	520	430	340	780	470	390	300	730	420	360	270	710	410	350	270	610	330	290	210
		3/4	-	-	-	500	1020	590	500	380	940	520	450	330	870	460	410	300	850	450	400	290	740	360	330	230
		7/8	-	-	-	540	1190	630	550	410	1090	550	500	360	1020	500	450	320	1000	490	440	310	860	390	370	250
		1	-	-	-	580	1360	680	610	440	1250	600	550	390	1160	540	500	350	1140	530	490	340	980	420	410	270
3	1-1/2	1/2	660	400	470	360	-	-	-	-	-	-	-	-	-	-	-	-	-	-	-	-	-	-	-	-
		5/8	940	560	550	460	-	-	-	-	-	-	-	-	-	-	-	-	-	-	-	-	-	-	-	-
		3/4	1270	660	620	500	-	-	-	-	-	-	-	-	-	-	-	-	-	-	-	-	-	-	-	-
		7/8	1520	720	690	540	-	-	-	-	-	-	-	-	-	-	-	-	-	-	-	-	-	-	-	-
		1	1740	770	750	580	-	-	-	-	-	-	-	-	-	-	-	-	-	-	-	-	-	-	-	-
3-1/8	1-1/2	1/2	-	-	-	-	610	370	430	330	580	340	390	310	550	320	360	290	540	320	340	280	490	280	280	230
		5/8	-	-	-	-	880	520	500	410	830	470	450	370	790	420	410	330	780	410	400	320	710	330	330	260
		3/4	-	-	-	-	1200	590	570	460	1130	520	510	410	1060	460	460	360	1040	450	450	350	890	360	370	280
		7/8	-	-	-	-	1440	630	630	490	1320	550	560	430	1230	500	510	390	1210	490	500	380	1040	390	410	310
		1	-	-	-	-	1640	680	690	530	1510	600	620	470	1410	540	560	420	1380	530	550	410	1190	420	450	330
5	1-1/2	5/8	940	560	640	500	-	-	-	-	-	-	-	-	-	-	-	-	-	-	-	-	-	-	-	-
		3/4	1270	660	850	660	-	-	-	-	-	-	-	-	-	-	-	-	-	-	-	-	-	-	-	-
		7/8	1680	720	1020	720	-	-	-	-	-	-	-	-	-	-	-	-	-	-	-	-	-	-	-	-
		1	2150	770	1100	770	-	-	-	-	-	-	-	-	-	-	-	-	-	-	-	-	-	-	-	-
5-1/8	1-1/2	5/8	-	-	-	-	880	520	590	460	830	470	560	430	790	420	530	410	780	410	520	400	710	330	460	330
		3/4	-	-	-	-	1200	590	790	590	1140	520	740	520	1100	460	670	460	1080	450	660	450	990	360	530	360
		7/8	-	-	-	-	1590	630	920	630	1520	550	810	550	1460	500	740	500	1440	490	720	490	1330	390	590	390
		1	-	-	-	-	2050	680	990	680	1930	600	890	600	1800	540	810	540	1760	530	780	530	1520	420	640	420
6-3/4	1-1/2	5/8	940	560	640	500	880	520	590	460	830	470	560	430	790	420	530	410	780	410	520	400	710	330	460	330
		3/4	1270	660	850	660	1200	590	790	590	1140	520	740	520	1100	460	700	460	1080	450	690	450	990	360	620	360
		7/8	1680	720	1090	720	1590	630	1010	630	1520	550	950	550	1460	500	900	500	1440	490	890	490	1330	390	750	390
		1	2150	770	1350	770	2050	680	1270	680	1930	600	1140	600	1800	540	1030	540	1760	530	1000	530	1520	420	810	420

1. Tabulated lateral design values (Z) for bolted connections shall be multiplied by all applicable adjustment factors (see Table 10.3.1).
2. Tabulated lateral design values (Z) are for "full diameter" (see Appendix L) bolts with bending yield strength (F_{yb}) of 45,000 psi.

Table 11D BOLTS: Reference Lateral Design Values (Z) for Single Shear (two member) Connections[1,2]

for structural glued laminated timber with 1/4" ASTM A 36 steel side plate

Main Member t_m in.	Side Member t_s in.	Bolt Diameter D in.	G=0.55 Southern Pine $Z_\parallel$ lbs.	$Z_\perp$ lbs.	G=0.50 Douglas Fir-Larch $Z_\parallel$ lbs.	$Z_\perp$ lbs.	G=0.46 Douglas Fir(S) Hem-Fir(N) $Z_\parallel$ lbs.	$Z_\perp$ lbs.	G=0.43 Hem-Fir $Z_\parallel$ lbs.	$Z_\perp$ lbs.	G=0.42 Spruce-Pine-Fir $Z_\parallel$ lbs.	$Z_\perp$ lbs.	G=0.36 Spruce-Pine-Fir(S) Western Woods $Z_\parallel$ lbs.	$Z_\perp$ lbs.
2-1/2	1/4	1/2	-	-	830	410	780	380	740	350	720	340	640	290
		5/8	-	-	1050	470	980	430	920	400	910	390	800	330
		3/4	-	-	1270	530	1180	490	1110	450	1090	440	960	370
		7/8	-	-	1480	590	1370	530	1290	490	1270	480	1120	410
		1	-	-	1690	640	1570	580	1480	540	1450	530	1280	450
3	1/4	1/2	860	540	-	-	-	-	-	-	-	-	-	-
		5/8	1260	610	-	-	-	-	-	-	-	-	-	-
		3/4	1610	670	-	-	-	-	-	-	-	-	-	-
		7/8	1880	740	-	-	-	-	-	-	-	-	-	-
		1	2150	790	-	-	-	-	-	-	-	-	-	-
3-1/8	1/4	1/2	-	-	830	490	800	440	770	410	770	400	720	330
		5/8	-	-	1210	550	1160	500	1110	460	1090	450	960	380
		3/4	-	-	1540	620	1420	560	1340	510	1310	500	1150	420
		7/8	-	-	1790	680	1660	610	1560	560	1530	550	1340	470
		1	-	-	2050	740	1900	670	1780	610	1750	600	1530	510
5	1/4	5/8	1260	760	-	-	-	-	-	-	-	-	-	-
		3/4	1740	1000	-	-	-	-	-	-	-	-	-	-
		7/8	2320	1140	-	-	-	-	-	-	-	-	-	-
		1	2980	1210	-	-	-	-	-	-	-	-	-	-
5-1/8	1/4	5/8	-	-	1210	710	1160	670	1130	640	1120	630	1050	550
		3/4	-	-	1670	940	1610	840	1560	760	1550	740	1450	610
		7/8	-	-	2220	1020	2140	900	2070	830	2050	810	1920	670
		1	-	-	2860	1100	2750	990	2670	900	2640	880	2390	720
6-3/4	1/4	5/8	1260	760	1210	710	1160	670	1130	640	1120	630	1050	570
		3/4	1740	1000	1670	940	1610	890	1560	850	1550	840	1450	750
		7/8	2320	1280	2220	1210	2140	1130	2070	1060	2050	1030	1920	850
		1	2980	1590	2860	1420	2750	1270	2670	1150	2640	1120	2470	910
8-1/2	1/4	3/4	1740	1000	-	-	-	-	-	-	-	-	-	-
		7/8	2320	1280	-	-	-	-	-	-	-	-	-	-
		1	2980	1590	-	-	-	-	-	-	-	-	-	-
8-3/4	1/4	3/4	-	-	1670	940	1610	890	1560	850	1550	840	1450	750
		7/8	-	-	2220	1210	2140	1130	2070	1080	2050	1070	1920	970
		1	-	-	2860	1500	2750	1420	2670	1350	2640	1330	2470	1150
10-1/2	1/4	7/8	2320	1280	-	-	-	-	-	-	-	-	-	-
		1	2980	1590	-	-	-	-	-	-	-	-	-	-
10-3/4	1/4	7/8	-	-	2220	1210	2140	1130	2070	1080	2050	1070	1920	970
		1	-	-	2860	1500	2750	1420	2670	1350	2640	1330	2470	1200
12-1/4	1/4	7/8	-	-	2220	1210	2140	1130	2070	1080	2050	1070	1920	970
		1	-	-	2860	1500	2750	1420	2670	1350	2640	1330	2470	1200
14-1/4	1/4	1	-	-	2860	1500	2750	1420	2670	1350	2640	1330	2470	1200

1. Tabulated lateral design values (Z) for bolted connections shall be multiplied by all applicable adjustment factors (see Table 10.3.1).
2. Tabulated lateral design values (Z) are for "full diameter" bolts (see Appendix L) with bending yield strength (F_{yb}) of 45,000 psi and dowel bearing strength (F_e) of 87,000 psi for ASTM A 36 steel.

Table 11E BOLTS: Reference Lateral Design Values (Z) for Single Shear (two member) Connections[1,2,3,4]

for sawn lumber or SCL to concrete

Embedment Depth in Concrete t_m in.	Side Member t_s in.	Bolt Diameter D in.	G=0.67 Red Oak $Z_{\parallel}$ lbs.	$Z_{\perp}$ lbs.	G=0.55 Mixed Maple Southern Pine $Z_{\parallel}$ lbs.	$Z_{\perp}$ lbs.	G=0.50 Douglas Fir-Larch $Z_{\parallel}$ lbs.	$Z_{\perp}$ lbs.	G=0.49 Douglas Fir-Larch(N) $Z_{\parallel}$ lbs.	$Z_{\perp}$ lbs.	G=0.46 Douglas Fir(S) Hem-Fir(N) $Z_{\parallel}$ lbs.	$Z_{\perp}$ lbs.
6.0 and greater	1-1/2	1/2	770	480	680	410	650	380	640	380	620	360
		5/8	1070	660	970	580	930	530	920	520	890	470
		3/4	1450	890	1330	660	1270	590	1260	560	1230	520
		7/8	1890	960	1750	720	1690	630	1680	600	1640	550
		1	2410	1020	2250	770	2100	680	2060	650	1930	600
	1-3/4	1/2	830	510	740	430	700	400	690	390	670	370
		5/8	1160	680	1030	600	980	550	970	550	940	530
		3/4	1530	900	1390	770	1330	680	1310	660	1270	600
		7/8	1970	1120	1800	840	1730	740	1720	700	1680	640
		1	2480	1190	2290	890	2210	790	2200	750	2150	700
	2-1/2	1/2	830	590	790	520	770	470	760	460	750	440
		5/8	1290	800	1230	670	1180	610	1170	610	1120	570
		3/4	1840	1000	1630	850	1540	800	1520	780	1460	750
		7/8	2290	1240	2050	1080	1940	1020	1920	1000	1860	920
		1	2800	1520	2530	1280	2410	1130	2390	1080	2310	1000
	3-1/2	1/2	830	590	790	540	770	510	760	500	750	490
		5/8	1290	880	1230	810	1200	730	1190	720	1170	670
		3/4	1860	1190	1770	980	1720	900	1720	880	1680	830
		7/8	2540	1410	2410	1190	2320	1100	2290	1070	2200	1020
		1	3310	1670	2970	1420	2800	1330	2770	1300	2660	1260

Embedment Depth in Concrete t_m in.	Side Member t_s in.	Bolt Diameter D in.	G=0.43 Hem-Fir $Z_{\parallel}$ lbs.	$Z_{\perp}$ lbs.	G=0.42 Spruce-Pine-Fir $Z_{\parallel}$ lbs.	$Z_{\perp}$ lbs.	G=0.37 Redwood (open grain) $Z_{\parallel}$ lbs.	$Z_{\perp}$ lbs.	G=0.36 Eastern Softwoods Spruce-Pine-Fir(S) Western Cedars Western Woods $Z_{\parallel}$ lbs.	$Z_{\perp}$ lbs.	G=0.35 Northern Species $Z_{\parallel}$ lbs.	$Z_{\perp}$ lbs.
6.0 and greater	1-1/2	1/2	590	340	590	340	550	310	540	290	530	290
		5/8	860	420	850	410	810	350	800	330	780	320
		3/4	1200	460	1190	450	1130	370	1120	360	1100	350
		7/8	1580	500	1540	490	1360	410	1330	390	1280	370
		1	1800	540	1760	530	1560	440	1520	420	1460	410
	1-3/4	1/2	640	360	630	350	580	320	580	310	560	310
		5/8	910	490	900	480	840	400	830	380	810	370
		3/4	1230	540	1220	530	1160	430	1140	420	1120	410
		7/8	1630	580	1610	570	1540	470	1520	460	1490	430
		1	2090	630	2060	610	1820	510	1770	490	1710	470
	2-1/2	1/2	730	410	730	400	700	360	690	340	680	340
		5/8	1070	540	1060	530	980	480	960	470	940	460
		3/4	1400	710	1380	700	1290	620	1270	600	1240	580
		7/8	1790	830	1770	810	1660	680	1640	660	1600	610
		1	2230	900	2210	880	2080	730	2060	700	2030	680
	3-1/2	1/2	730	470	730	470	700	430	690	410	690	400
		5/8	1140	620	1140	610	1090	550	1080	530	1070	520
		3/4	1650	780	1640	770	1540	680	1510	670	1470	660
		7/8	2100	960	2070	950	1910	870	1880	850	1840	820
		1	2550	1190	2520	1180	2340	1020	2310	980	2260	950

1. Tabulated lateral design values (Z) for bolted connections shall be multiplied by all applicable adjustment factors (see Table 10.3.1).
2. Tabulated lateral design values (Z) are for "full diameter" bolts (see Appendix L) with bending yield strength (F_{yb}) of 45,000 psi.
3. Tabulated lateral design values (Z) are based on dowel bearing strength (F_e) of 7,500 psi for concrete with minimum $f_c' = 2,500$ psi.
4. Six inch anchor embedment assumed.

Table 11F BOLTS: Reference Lateral Design Values (Z) for Double Shear (three member) Connections[1,2]

for sawn lumber or SCL with all members of identical specific gravity

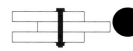

Thickness			G=0.67 Red Oak			G=0.55 Mixed Maple Southern Pine			G=0.50 Douglas Fir-Larch			G=0.49 Douglas Fir-Larch(N)			G=0.46 Douglas Fir(S) Hem-Fir(N)		
Main Member t_m in.	Side Member t_s in.	Bolt Diameter D in.	$Z_{\parallel}$ lbs.	$Z_{s\perp}$ lbs.	$Z_{m\perp}$ lbs.	$Z_{\parallel}$ lbs.	$Z_{s\perp}$ lbs.	$Z_{m\perp}$ lbs.	$Z_{\parallel}$ lbs.	$Z_{s\perp}$ lbs.	$Z_{m\perp}$ lbs.	$Z_{\parallel}$ lbs.	$Z_{s\perp}$ lbs.	$Z_{m\perp}$ lbs.	$Z_{\parallel}$ lbs.	$Z_{s\perp}$ lbs.	$Z_{m\perp}$ lbs.
1-1/2	1-1/2	1/2	1410	960	730	1150	800	550	1050	730	470	1030	720	460	970	680	420
		5/8	1760	1310	810	1440	1130	610	1310	1040	530	1290	1030	520	1210	940	470
		3/4	2110	1690	890	1730	1330	660	1580	1170	590	1550	1130	560	1450	1040	520
		7/8	2460	1920	960	2020	1440	720	1840	1260	630	1800	1210	600	1690	1100	550
		1	2810	2040	1020	2310	1530	770	2100	1350	680	2060	1290	650	1930	1200	600
1-3/4	1-3/4	1/2	1640	1030	850	1350	850	640	1230	770	550	1200	750	530	1130	710	490
		5/8	2050	1370	940	1680	1160	710	1530	1070	610	1500	1060	600	1410	1000	550
		3/4	2460	1810	1040	2020	1550	770	1840	1370	680	1800	1310	660	1690	1210	600
		7/8	2870	2240	1120	2350	1680	840	2140	1470	740	2110	1410	700	1970	1290	640
		1	3280	2380	1190	2690	1790	890	2450	1580	790	2410	1510	750	2250	1400	700
2-1/2	1-1/2	1/2	1530	960	1120	1320	800	910	1230	730	790	1210	720	760	1160	680	700
		5/8	2150	1310	1340	1870	1130	1020	1760	1040	880	1740	1030	860	1660	940	780
		3/4	2890	1770	1480	2550	1330	1110	2400	1170	980	2380	1130	940	2280	1040	860
		7/8	3780	1920	1600	3360	1440	1200	3060	1260	1050	3010	1210	1010	2820	1100	920
		1	4690	2040	1700	3840	1530	1280	3500	1350	1130	3440	1290	1080	3220	1200	1000
	1-1/2	1/2	1530	960	1120	1320	800	940	1230	730	860	1210	720	850	1160	680	810
		5/8	2150	1310	1510	1870	1130	1290	1760	1040	1190	1740	1030	1170	1660	940	1090
		3/4	2890	1770	1980	2550	1330	1550	2400	1170	1370	2380	1130	1310	2280	1040	1210
		7/8	3780	1920	2240	3360	1440	1680	3180	1260	1470	3150	1210	1410	3030	1100	1290
		1	4820	2040	2380	4310	1530	1790	4090	1350	1580	4050	1290	1510	3860	1200	1400
3-1/2	1-3/4	1/2	1660	1030	1180	1430	850	1030	1330	770	940	1310	750	920	1250	710	870
		5/8	2310	1370	1630	1990	1160	1380	1860	1070	1230	1840	1060	1200	1760	1000	1090
		3/4	3060	1810	2070	2670	1550	1550	2510	1370	1370	2480	1310	1310	2370	1210	1210
		7/8	3940	2240	2240	3470	1680	1680	3270	1470	1470	3240	1410	1410	3110	1290	1290
		1	4960	2380	2380	4400	1790	1790	4170	1580	1580	4120	1510	1510	3970	1400	1400
	3-1/2	1/2	1660	1180	1180	1500	1040	1040	1430	970	970	1420	960	960	1370	920	920
		5/8	2590	1770	1770	2340	1560	1420	2240	1410	1230	2220	1390	1200	2150	1290	1090
		3/4	3730	2380	2070	3380	1910	1550	3220	1750	1370	3190	1700	1310	3090	1610	1210
		7/8	5080	2820	2240	4600	2330	1680	4290	2130	1470	4210	2070	1410	3940	1960	1290
		1	6560	3340	2380	5380	2780	1790	4900	2580	1580	4810	2520	1510	4510	2410	1400
5-1/4	1-1/2	5/8	2150	1310	1510	1870	1130	1290	1760	1040	1190	1740	1030	1170	1660	940	1110
		3/4	2890	1770	1980	2550	1330	1690	2400	1170	1580	2380	1130	1550	2280	1040	1480
		7/8	3780	1920	2520	3360	1440	2170	3180	1260	2030	3150	1210	1990	3030	1100	1900
		1	4820	2040	3120	4310	1530	2680	4090	1350	2360	4050	1290	2260	3860	1200	2100
	1-3/4	5/8	2310	1370	1630	1990	1160	1380	1860	1070	1270	1840	1060	1250	1760	1000	1180
		3/4	3060	1810	2110	2670	1550	1790	2510	1370	1660	2480	1310	1630	2370	1210	1550
		7/8	3940	2240	2640	3470	1680	2260	3270	1470	2100	3240	1410	2060	3110	1290	1930
		1	4960	2380	3240	4400	1790	2680	4170	1580	2360	4120	1510	2260	3970	1400	2100
	3-1/2	5/8	2590	1770	1770	2340	1560	1560	2240	1410	1460	2220	1390	1450	2150	1290	1390
		3/4	3730	2380	2480	3380	1910	2180	3220	1750	2050	3190	1700	1970	3090	1610	1810
		7/8	5080	2820	3290	4600	2330	2530	4390	2130	2210	4350	2070	2110	4130	1960	1930
		1	6630	3340	3570	5740	2780	2680	5330	2580	2360	5250	2520	2260	4990	2410	2100
5-1/2	1-1/2	5/8	2150	1310	1510	1870	1130	1290	1760	1040	1190	1740	1030	1170	1660	940	1110
		3/4	2890	1770	1980	2550	1330	1690	2400	1170	1580	2380	1130	1550	2280	1040	1480
		7/8	3780	1920	2520	3360	1440	2170	3180	1260	2030	3150	1210	1990	3030	1100	1900
		1	4820	2040	3120	4310	1530	2700	4090	1350	2480	4050	1290	2370	3860	1200	2200
	3-1/2	5/8	2590	1770	1770	2340	1560	1560	2240	1410	1460	2220	1390	1450	2150	1290	1390
		3/4	3730	2380	2480	3380	1910	2180	3220	1750	2050	3190	1700	2020	3090	1610	1900
		7/8	5080	2820	3290	4600	2330	2650	4390	2130	2310	4350	2070	2210	4130	1960	2020
		1	6630	3340	3740	5740	2780	2810	5330	2580	2480	5250	2520	2370	4990	2410	2200
7-1/2	1-1/2	5/8	2150	1310	1510	1870	1130	1290	1760	1040	1190	1740	1030	1170	1660	940	1110
		3/4	2890	1770	1980	2550	1330	1690	2400	1170	1580	2380	1130	1550	2280	1040	1480
		7/8	3780	1920	2520	3360	1440	2170	3180	1260	2030	3150	1210	1990	3030	1100	1900
		1	4820	2040	3120	4310	1530	2700	4090	1350	2530	4050	1290	2480	3860	1200	2390
	3-1/2	5/8	2590	1770	1770	2340	1560	1560	2240	1410	1460	2220	1390	1450	2150	1290	1390
		3/4	3730	2380	2480	3380	1910	2180	3220	1750	2050	3190	1700	2020	3090	1610	1940
		7/8	5080	2820	3290	4600	2330	2890	4390	2130	2720	4350	2070	2670	4130	1960	2560
		1	6630	3340	4190	5740	2780	3680	5330	2580	3380	5250	2520	3230	4990	2410	3000

1. Tabulated lateral design values (Z) for bolted connections shall be multiplied by all applicable adjustment factors (see Table 10.3.1).
2. Tabulated lateral design values (Z) are for "full diameter" bolts (see Appendix L) with bending yield strength (F_{yb}) of 45,000 psi.

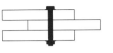

Table 11F (Cont.)   BOLTS: Reference Lateral Design Values (Z) for Double Shear (three member) Connections[1,2]

for sawn lumber or SCL with all members of identical specific gravity

Main Member t_m (in.)	Side Member t_s (in.)	Bolt Diameter D (in.)	G=0.43 Hem-Fir $Z_{\parallel}$ lbs.	$Z_{s\perp}$ lbs.	$Z_{m\perp}$ lbs.	G=0.42 Spruce-Pine-Fir $Z_{\parallel}$ lbs.	$Z_{s\perp}$ lbs.	$Z_{m\perp}$ lbs.	G=0.37 Redwood (open grain) $Z_{\parallel}$ lbs.	$Z_{s\perp}$ lbs.	$Z_{m\perp}$ lbs.	G=0.36 Eastern Softwoods Spruce-Pine-Fir(S) Western Cedars Western Woods $Z_{\parallel}$ lbs.	$Z_{s\perp}$ lbs.	$Z_{m\perp}$ lbs.	G=0.35 Northern Species $Z_{\parallel}$ lbs.	$Z_{s\perp}$ lbs.	$Z_{m\perp}$ lbs.
1-1/2	1-1/2	1/2	900	650	380	880	640	370	780	580	310	760	560	290	730	550	290
		5/8	1130	840	420	1100	830	410	970	690	350	950	660	330	910	640	320
		3/4	1350	920	460	1320	900	450	1170	740	370	1140	720	360	1100	700	350
		7/8	1580	1000	500	1540	970	490	1360	810	410	1330	790	390	1280	740	370
		1	1800	1080	540	1760	1050	530	1560	870	440	1520	840	420	1460	810	410
1-3/4	1-3/4	1/2	1050	670	450	1030	660	430	910	590	360	890	580	340	850	570	330
		5/8	1310	950	490	1290	940	480	1130	810	400	1110	770	380	1070	740	370
		3/4	1580	1080	540	1540	1050	530	1360	870	430	1330	840	420	1280	810	410
		7/8	1840	1160	580	1800	1130	570	1590	950	470	1550	920	460	1490	860	430
		1	2100	1260	630	2060	1230	610	1820	1020	510	1770	980	490	1710	950	470
2-1/2	1-1/2	1/2	1100	650	640	1080	640	610	990	580	510	980	560	490	950	550	480
		5/8	1590	840	700	1570	830	690	1450	690	580	1430	660	550	1390	640	530
		3/4	2190	920	770	2160	900	750	1950	740	620	1900	720	600	1830	700	580
		7/8	2630	1000	830	2570	970	810	2270	810	680	2210	790	660	2130	740	610
		1	3000	1080	900	2940	1050	880	2590	870	730	2530	840	700	2440	810	680
3-1/2	1-1/2	1/2	1100	650	760	1080	640	740	990	580	670	980	560	660	950	550	640
		5/8	1590	840	980	1570	830	960	1450	690	810	1430	660	770	1390	640	740
		3/4	2190	920	1080	2160	900	1050	2010	740	870	1990	720	840	1940	700	810
		7/8	2920	1000	1160	2880	970	1130	2690	810	950	2660	790	920	2560	740	860
		1	3600	1080	1260	3530	1050	1230	3110	870	1020	3040	840	980	2930	810	950
3-1/2	1-3/4	1/2	1180	670	820	1160	660	800	1060	590	720	1040	580	680	1010	570	670
		5/8	1670	950	980	1650	940	960	1510	810	810	1490	770	770	1450	740	740
		3/4	2270	1080	1080	2240	1050	1050	2070	870	870	2040	840	840	1990	810	810
		7/8	2980	1160	1160	2950	1130	1130	2740	950	950	2700	920	920	2640	860	860
		1	3820	1260	1260	3770	1230	1230	3520	1020	1020	3480	980	980	3410	950	950
3-1/2	3-1/2	1/2	1330	880	880	1310	870	860	1230	800	720	1220	780	680	1200	760	670
		5/8	2070	1190	980	2050	1170	960	1930	1030	810	1900	1000	770	1870	970	740
		3/4	2980	1490	1080	2950	1460	1050	2720	1290	870	2660	1270	840	2560	1240	810
		7/8	3680	1840	1160	3600	1810	1130	3180	1640	950	3100	1610	920	2990	1550	860
		1	4200	2280	1260	4110	2240	1230	3630	2030	1020	3540	1960	980	3410	1890	950
5-1/4	1-1/2	5/8	1590	840	1050	1570	830	1040	1450	690	940	1430	660	920	1390	640	900
		3/4	2190	920	1400	2160	900	1380	2010	740	1250	1990	720	1230	1940	700	1210
		7/8	2920	1000	1750	2880	970	1700	2690	810	1420	2660	790	1380	2560	740	1290
		1	3600	1080	1890	3530	1050	1840	3110	870	1520	3040	840	1470	2930	810	1420
5-1/4	1-3/4	5/8	1670	950	1110	1650	940	1100	1510	810	990	1490	770	970	1450	740	940
		3/4	2270	1080	1460	2240	1050	1440	2070	870	1300	2040	840	1260	1990	810	1220
		7/8	2980	1160	1750	2950	1130	1700	2740	950	1420	2700	920	1380	2640	860	1290
		1	3820	1260	1890	3770	1230	1840	3520	1020	1520	3480	980	1470	3410	950	1420
5-1/4	3-1/2	5/8	2070	1190	1320	2050	1170	1310	1930	1030	1210	1900	1000	1150	1870	970	1120
		3/4	2980	1490	1610	2950	1460	1580	2770	1290	1300	2740	1270	1260	2660	1240	1220
		7/8	3900	1840	1750	3840	1810	1700	3480	1640	1420	3410	1610	1380	3320	1550	1290
		1	4730	2280	1890	4660	2240	1840	4240	2030	1520	4170	1960	1470	4050	1890	1420
5-1/2	1-1/2	5/8	1590	840	1050	1570	830	1040	1450	690	940	1430	660	920	1390	640	900
		3/4	2190	920	1400	2160	900	1380	2010	740	1250	1990	720	1230	1940	700	1210
		7/8	2920	1000	1800	2880	970	1780	2690	810	1490	2660	790	1440	2560	740	1350
		1	3600	1080	1980	3530	1050	1930	3110	870	1600	3040	840	1540	2930	810	1490
5-1/2	3-1/2	5/8	2070	1190	1320	2050	1170	1310	1930	1030	1210	1900	1000	1180	1870	970	1160
		3/4	2980	1490	1690	2950	1460	1650	2770	1290	1360	2740	1270	1320	2660	1240	1280
		7/8	3900	1840	1830	3840	1810	1780	3480	1640	1490	3410	1610	1440	3320	1550	1350
		1	4730	2280	1980	4660	2240	1930	4240	2030	1600	4170	1960	1540	4050	1890	1490
7-1/2	1-1/2	5/8	1590	840	1050	1570	830	1040	1450	690	940	1430	660	920	1390	640	900
		3/4	2190	920	1400	2160	900	1380	2010	740	1250	1990	720	1230	1940	700	1210
		7/8	2920	1000	1800	2880	970	1780	2690	810	1630	2660	790	1600	2560	740	1550
		1	3600	1080	2270	3530	1050	2240	3110	870	2040	3040	840	2010	2930	810	1970
7-1/2	3-1/2	5/8	2070	1190	1320	2050	1170	1310	1930	1030	1210	1900	1000	1180	1870	970	1160
		3/4	2980	1490	1850	2950	1460	1820	2770	1290	1670	2740	1270	1650	2660	1240	1620
		7/8	3900	1840	2450	3840	1810	2420	3480	1640	2030	3410	1610	1970	3320	1550	1840
		1	4730	2280	2700	4660	2240	2630	4240	2030	2180	4170	1960	2100	4050	1890	2030

1. Tabulated lateral design values (Z) for bolted connections shall be multiplied by all applicable adjustment factors (see Table 10.3.1).
2. Tabulated lateral design values (Z) are for "full diameter" bolts (see Appendix L) with bending yield strength (F_{yb}) of 45,000 psi.

BOLTS

DOWEL-TYPE FASTENERS

11

BOLTS

Table 11G BOLTS: Reference Lateral Design Values (Z) for Double Shear (three member) Connections[1,2]

for sawn lumber or SCL with 1/4" ASTM A 36 steel side plate

Main Member t_m in.	Side Member t_s in.	Bolt Diameter D in.	G=0.67 Red Oak $Z_\parallel$ lbs.	$Z_\perp$ lbs.	G=0.55 Mixed Maple Southern Pine $Z_\parallel$ lbs.	$Z_\perp$ lbs.	G=0.50 Douglas Fir-Larch $Z_\parallel$ lbs.	$Z_\perp$ lbs.	G=0.49 Douglas Fir-Larch (N) $Z_\parallel$ lbs.	$Z_\perp$ lbs.	G=0.46 Douglas Fir(S) Hem-Fir(N) $Z_\parallel$ lbs.	$Z_\perp$ lbs.	G=0.43 Hem-Fir $Z_\parallel$ lbs.	$Z_\perp$ lbs.	G=0.42 Spruce-Pine-Fir $Z_\parallel$ lbs.	$Z_\perp$ lbs.	G=0.37 Redwood (open grain) $Z_\parallel$ lbs.	$Z_\perp$ lbs.	G=0.36 Eastern Softwoods Spruce-Pine-Fir(S) Western Cedars Western Woods $Z_\parallel$ lbs.	$Z_\perp$ lbs.	G=0.35 Northern Species $Z_\parallel$ lbs.	$Z_\perp$ lbs.
1-1/2	1/4	1/2	1410	730	1150	550	1050	470	1030	460	970	420	900	380	880	370	780	310	760	290	730	290
		5/8	1760	810	1440	610	1310	530	1290	520	1210	470	1130	420	1100	410	970	350	950	330	910	320
		3/4	2110	890	1730	660	1580	590	1550	560	1450	520	1350	460	1320	450	1170	370	1140	360	1100	350
		7/8	2460	960	2020	720	1840	630	1800	600	1690	550	1580	500	1540	490	1360	410	1330	390	1280	370
		1	2810	1020	2310	770	2100	680	2060	650	1930	600	1800	540	1760	530	1560	440	1520	420	1460	410
1-3/4	1/4	1/2	1640	850	1350	640	1230	550	1200	530	1130	490	1050	450	1030	430	910	360	890	340	850	330
		5/8	2050	940	1680	710	1530	610	1500	600	1410	550	1310	490	1290	480	1130	400	1110	380	1070	370
		3/4	2460	1040	2020	770	1840	680	1800	660	1690	600	1580	540	1540	530	1360	430	1330	420	1280	410
		7/8	2870	1120	2350	840	2140	740	2110	700	1970	640	1840	580	1800	570	1590	470	1550	460	1490	430
		1	3280	1190	2690	890	2450	790	2410	750	2250	700	2100	630	2060	610	1820	510	1770	490	1710	470
2-1/2	1/4	1/2	1870	1210	1720	910	1650	790	1640	760	1590	700	1500	640	1470	610	1300	510	1270	490	1220	480
		5/8	2740	1340	2400	1020	2190	880	2150	860	2010	780	1880	700	1840	690	1620	580	1580	550	1520	530
		3/4	3520	1480	2880	1110	2630	980	2580	940	2410	860	2250	770	2200	750	1950	620	1900	600	1830	580
		7/8	4100	1600	3360	1200	3060	1050	3010	1010	2820	920	2630	830	2570	810	2270	680	2210	660	2130	610
		1	4690	1700	3840	1280	3500	1130	3440	1080	3220	1000	3000	900	2940	880	2590	730	2530	700	2440	680
3-1/2	1/4	1/2	1870	1240	1720	1100	1650	1030	1640	1010	1590	970	1540	890	1530	860	1450	720	1430	680	1410	670
		5/8	2740	1720	2510	1420	2410	1230	2390	1200	2330	1090	2260	980	2230	960	2110	810	2090	770	2060	740
		3/4	3800	2070	3480	1550	3340	1370	3320	1310	3220	1210	3120	1080	3080	1050	2720	870	2660	840	2560	810
		7/8	5060	2240	4630	1680	4290	1470	4210	1410	3940	1290	3680	1160	3600	1130	3180	950	3100	920	2990	860
		1	6520	2380	5380	1790	4900	1580	4810	1510	4510	1400	4200	1260	4110	1230	3630	1020	3540	980	3410	950
5-1/4	1/4	5/8	2740	1720	2510	1510	2410	1420	2390	1400	2330	1340	2260	1280	2230	1270	2110	1170	2090	1140	2060	1120
		3/4	3800	2290	3480	2000	3340	1890	3320	1850	3220	1780	3120	1610	3090	1580	2920	1300	2890	1260	2840	1220
		7/8	5060	2930	4630	2530	4440	2210	4410	2110	4280	1930	4150	1750	4110	1700	3880	1420	3840	1380	3770	1290
		1	6520	3570	5960	2680	5720	2360	5670	2260	5510	2100	5330	1890	5280	1840	4990	1520	4930	1470	4850	1420
5-1/2	1/4	5/8	2740	1720	2510	1510	2410	1420	2390	1400	2330	1340	2260	1280	2230	1270	2110	1170	2090	1140	2060	1120
		3/4	3800	2290	3480	2000	3340	1890	3320	1850	3220	1780	3120	1690	3090	1650	2920	1360	2890	1320	2840	1280
		7/8	5060	2930	4630	2570	4440	2310	4410	2210	4280	2020	4150	1830	4110	1780	3880	1490	3840	1440	3770	1350
		1	6520	3640	5960	2810	5720	2480	5670	2370	5510	2200	5330	1980	5280	1930	4990	1600	4930	1540	4850	1490
7-1/2	1/4	5/8	2740	1720	2510	1510	2410	1420	2390	1400	2330	1340	2260	1280	2230	1270	2110	1170	2090	1140	2060	1120
		3/4	3800	2290	3480	2000	3340	1890	3320	1850	3220	1780	3120	1690	3090	1670	2920	1530	2890	1500	2840	1480
		7/8	5060	2930	4630	2570	4440	2410	4410	2360	4280	2260	4150	2160	4110	2130	3880	1960	3840	1930	3770	1840
		1	6520	3640	5960	3180	5720	3000	5670	2940	5510	2840	5330	2700	5280	2630	4990	2180	4930	2100	4850	2030
9-1/2	1/4	3/4	3800	2290	3480	2000	3340	1890	3320	1850	3220	1780	3120	1690	3090	1670	2920	1530	2890	1500	2840	1480
		7/8	5060	2930	4630	2570	4440	2410	4410	2360	4280	2260	4150	2160	4110	2130	3880	1960	3840	1930	3770	1870
		1	6520	3640	5960	3180	5720	3000	5670	2940	5510	2840	5330	2700	5280	2660	4990	2440	4930	2400	4850	2350
11-1/2	1/4	7/8	5060	2930	4630	2570	4440	2410	4410	2360	4280	2260	4150	2160	4110	2130	3880	1960	3840	1930	3770	1870
		1	6520	3640	5960	3180	5720	3000	5670	2940	5510	2840	5330	2700	5280	2660	4990	2440	4930	2400	4850	2350
13-1/2	1/4	1	6520	3640	5960	3180	5720	3000	5670	2940	5510	2840	5330	2700	5280	2660	4990	2440	4930	2400	4850	2350

1. Tabulated lateral design values (Z) for bolted connections shall be multiplied by all applicable adjustment factors (see Table 10.3.1).

2. Tabulated lateral design values (Z) are for "full diameter" bolts (see Appendix L) with bending yield strength (F_{yb}) of 45,000 psi and a dowel bearing strength (F_e) of 87,000 psi for ASTM A 36 steel.

Table 11H BOLTS: Reference Lateral Design Values (Z) for Double Shear (three member) Connections[1,2]

for structural glued laminated timber main member with sawn lumber side member of identical species

Main Member t_m in.	Side Member t_s in.	Bolt Diameter D in.	G=0.55 Southern Pine $Z_{\parallel}$ lbs.	$Z_{s\perp}$ lbs.	$Z_{m\perp}$ lbs.	G=0.50 Douglas Fir-Larch $Z_{\parallel}$ lbs.	$Z_{s\perp}$ lbs.	$Z_{m\perp}$ lbs.	G=0.46 Douglas Fir(S) Hem-Fir(N) $Z_{\parallel}$ lbs.	$Z_{s\perp}$ lbs.	$Z_{m\perp}$ lbs.	G=0.43 Hem-Fir $Z_{\parallel}$ lbs.	$Z_{s\perp}$ lbs.	$Z_{m\perp}$ lbs.	G=0.42 Spruce-Pine-Fir $Z_{\parallel}$ lbs.	$Z_{s\perp}$ lbs.	$Z_{m\perp}$ lbs.	G=0.36 Spruce-Pine-Fir(S) Western Woods $Z_{\parallel}$ lbs.	$Z_{s\perp}$ lbs.	$Z_{m\perp}$ lbs.
2-1/2	1-1/2	1/2	-	-	-	1230	730	790	1160	680	700	1100	650	640	1080	640	610	980	560	490
		5/8	-	-	-	1760	1040	880	1660	940	780	1590	840	700	1570	830	690	1430	660	550
		3/4	-	-	-	2400	1170	980	2280	1040	860	2190	920	770	2160	900	750	1900	720	600
		7/8	-	-	-	3060	1260	1050	2820	1100	920	2630	1000	830	2570	970	810	2210	790	660
		1	-	-	-	3500	1350	1130	3220	1200	1000	3000	1080	900	2940	1050	880	2530	840	700
3	1-1/2	1/2	1320	800	940	-	-	-	-	-	-	-	-	-	-	-	-	-	-	-
		5/8	1870	1130	1220	-	-	-	-	-	-	-	-	-	-	-	-	-	-	-
		3/4	2550	1330	1330	-	-	-	-	-	-	-	-	-	-	-	-	-	-	-
		7/8	3360	1440	1440	-	-	-	-	-	-	-	-	-	-	-	-	-	-	-
		1	4310	1530	1530	-	-	-	-	-	-	-	-	-	-	-	-	-	-	-
3-1/8	1-1/2	1/2	-	-	-	1230	730	860	1160	680	810	1100	650	760	1080	640	740	980	560	610
		5/8	-	-	-	1760	1040	1090	1660	940	980	1590	840	880	1570	830	860	1430	660	680
		3/4	-	-	-	2400	1170	1220	2280	1040	1080	2190	920	960	2160	900	940	1990	720	750
		7/8	-	-	-	3180	1260	1310	3030	1100	1150	2920	1000	1040	2880	970	1010	2660	790	820
		1	-	-	-	4090	1350	1410	3860	1200	1250	3600	1080	1130	3530	1050	1090	3040	840	880
5	1-1/2	5/8	1870	1130	1290	-	-	-	-	-	-	-	-	-	-	-	-	-	-	-
		3/4	2550	1330	1690	-	-	-	-	-	-	-	-	-	-	-	-	-	-	-
		7/8	3360	1440	2170	-	-	-	-	-	-	-	-	-	-	-	-	-	-	-
		1	4310	1530	2550	-	-	-	-	-	-	-	-	-	-	-	-	-	-	-
5-1/8	1-1/2	5/8	-	-	-	1760	1040	1190	1660	940	1110	1590	840	1050	1570	830	1040	1430	660	920
		3/4	-	-	-	2400	1170	1580	2280	1040	1480	2190	920	1400	2160	900	1380	1990	720	1230
		7/8	-	-	-	3180	1260	2030	3030	1100	1880	2920	1000	1700	2880	970	1660	2660	790	1350
		1	-	-	-	4090	1350	2310	3860	1200	2050	3600	1080	1850	3530	1050	1790	3040	840	1440
6-3/4	1-1/2	5/8	1870	1130	1290	1760	1040	1190	1660	940	1110	1590	840	1050	1570	830	1040	1430	660	920
		3/4	2550	1330	1690	2400	1170	1580	2280	1040	1480	2190	920	1400	2160	900	1380	1990	720	1230
		7/8	3360	1440	2170	3180	1260	2030	3030	1100	1900	2920	1000	1800	2880	970	1780	2660	790	1600
		1	4310	1530	2700	4090	1350	2530	3860	1200	2390	3600	1080	2270	3530	1050	2240	3040	840	1890

1. Tabulated lateral design values (Z) for bolted connections shall be multiplied by all applicable adjustment factors (see Table 10.3.1).
2. Tabulated lateral design values (Z) are for "full diameter" bolts (see Appendix L) with bending yield strength (F_{yb}) of 45,000 psi.

BOLTS

DOWEL-TYPE FASTENERS

11

BOLTS

Table 11I   BOLTS: Reference Lateral Design Values (Z) for Double Shear (three member) Connections[1,2]

for structural glued laminated timber with 1/4" ASTM A 36 steel side plate

Main Member t_m in.	Side Member t_s in.	Bolt Diameter D in.	G=0.55 Southern Pine $Z_{\parallel}$ lbs.	$Z_{\perp}$ lbs.	G=0.50 Douglas Fir-Larch $Z_{\parallel}$ lbs.	$Z_{\perp}$ lbs.	G=0.46 Douglas Fir(S) Hem-Fir(N) $Z_{\parallel}$ lbs.	$Z_{\perp}$ lbs.	G=0.43 Hem-Fir $Z_{\parallel}$ lbs.	$Z_{\perp}$ lbs.	G=0.42 Spruce-Pine-Fir $Z_{\parallel}$ lbs.	$Z_{\perp}$ lbs.	G=0.36 Spruce-Pine-Fir(S) Western Woods $Z_{\parallel}$ lbs.	$Z_{\perp}$ lbs.
2-1/2	1/4	1/2	-	-	1650	790	1590	700	1500	640	1470	610	1270	490
		5/8	-	-	2190	880	2010	780	1880	700	1840	690	1580	550
		3/4	-	-	2630	980	2410	860	2250	770	2200	750	1900	600
		7/8	-	-	3060	1050	2820	920	2630	830	2570	810	2210	660
		1	-	-	3500	1130	3220	1000	3000	900	2940	880	2530	700
3	1/4	1/2	1720	1100	-	-	-	-	-	-	-	-	-	-
		5/8	2510	1220	-	-	-	-	-	-	-	-	-	-
		3/4	3460	1330	-	-	-	-	-	-	-	-	-	-
		7/8	4040	1440	-	-	-	-	-	-	-	-	-	-
		1	4610	1530	-	-	-	-	-	-	-	-	-	-
3-1/8	1/4	1/2	-	-	1650	980	1590	880	1540	800	1530	770	1430	610
		5/8	-	-	2410	1090	2330	980	2260	880	2230	860	1980	680
		3/4	-	-	3280	1220	3020	1080	2810	960	2750	940	2370	750
		7/8	-	-	3830	1310	3520	1150	3280	1040	3210	1010	2770	820
		1	-	-	4380	1410	4020	1250	3750	1130	3670	1090	3160	880
5	1/4	5/8	2510	1510	-	-	-	-	-	-	-	-	-	-
		3/4	3480	2000	-	-	-	-	-	-	-	-	-	-
		7/8	4630	2410	-	-	-	-	-	-	-	-	-	-
		1	5960	2550	-	-	-	-	-	-	-	-	-	-
5-1/8	1/4	5/8	-	-	2410	1420	2330	1340	2260	1280	2230	1270	2090	1120
		3/4	-	-	3340	1890	3220	1770	3120	1580	3090	1540	2890	1230
		7/8	-	-	4440	2150	4280	1880	4150	1700	4110	1660	3840	1350
		1	-	-	5720	2310	5510	2050	5330	1850	5280	1790	4930	1440
6-3/4	1/4	5/8	2510	1510	2410	1420	2330	1340	2260	1280	2230	1270	2090	1140
		3/4	3480	2000	3340	1890	3220	1780	3120	1690	3090	1670	2890	1500
		7/8	4630	2570	4440	2410	4280	2260	4150	2160	4110	2130	3840	1770
		1	5960	3180	5720	3000	5510	2700	5330	2430	5280	2360	4930	1890
8-1/2	1/4	3/4	3480	2000	-	-	-	-	-	-	-	-	-	-
		7/8	4630	2570	-	-	-	-	-	-	-	-	-	-
		1	5960	3180	-	-	-	-	-	-	-	-	-	-
8-3/4	1/4	3/4	-	-	3340	1890	3220	1780	3120	1690	3090	1670	2890	1500
		7/8	-	-	4440	2410	4280	2260	4150	2160	4110	2130	3840	1930
		1	-	-	5720	3000	5510	2840	5330	2700	5280	2660	4930	2400
10-1/2	1/4	7/8	4630	2570	-	-	-	-	-	-	-	-	-	-
		1	5960	3180	-	-	-	-	-	-	-	-	-	-
10-3/4	1/4	7/8	-	-	4440	2410	4280	2260	4150	2160	4110	2130	3840	1930
		1	-	-	5720	3000	5510	2840	5330	2700	5280	2660	4930	2400
12-1/4	1/4	7/8	-	-	4440	2410	4280	2260	4150	2160	4110	2130	3840	1930
		1	-	-	5720	3000	5510	2840	5330	2700	5280	2660	4930	2400
14-1/4	1/4	1	-	-	5720	3000	5510	2840	5330	2700	5280	2660	4930	2400

1. Tabulated lateral design values (Z) for bolted connections shall be multiplied by all applicable adjustment factors (see Table 10.3.1).
2. Tabulated lateral design values (Z) are for "full diameter" bolts (see Appendix L) with bending yield strength (F_{yb}) of 45,000 psi and dowel bearing strength (F_e) of 87,000 psi for ASTM A 36 steel.

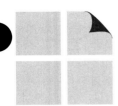

This page left blank intentionally.

DOWEL-TYPE FASTENERS

11

Table 11J LAG SCREWS: Reference Lateral Design Values (Z) for Single Shear (two member) Connections[1,2,3]

for sawn lumber or SCL with both members of identical specific gravity

Side Member Thickness t_s (in.)	Lag Screw Diameter D (in.)	G=0.67 Red Oak $Z_\parallel$ (lbs.)	$Z_{s\perp}$ (lbs.)	$Z_{m\perp}$ (lbs.)	$Z_\perp$ (lbs.)	G=0.55 Mixed Maple Southern Pine $Z_\parallel$ (lbs.)	$Z_{s\perp}$ (lbs.)	$Z_{m\perp}$ (lbs.)	$Z_\perp$ (lbs.)	G=0.50 Douglas Fir-Larch $Z_\parallel$ (lbs.)	$Z_{s\perp}$ (lbs.)	$Z_{m\perp}$ (lbs.)	$Z_\perp$ (lbs.)	G=0.49 Douglas Fir-Larch(N) $Z_\parallel$ (lbs.)	$Z_{s\perp}$ (lbs.)	$Z_{m\perp}$ (lbs.)	$Z_\perp$ (lbs.)	G=0.46 Douglas Fir(S) Hem-Fir(N) $Z_\parallel$ (lbs.)	$Z_{s\perp}$ (lbs.)	$Z_{m\perp}$ (lbs.)	$Z_\perp$ (lbs.)
1/2	1/4	150	110	110	110	130	90	100	90	120	90	90	80	120	90	90	80	110	80	90	80
	5/16	170	130	130	120	150	110	120	100	150	100	110	100	140	100	110	90	140	100	100	90
	3/8	180	130	130	120	160	110	110	100	150	100	110	90	150	90	110	90	140	90	100	90
5/8	1/4	160	120	130	120	140	100	110	100	130	90	100	90	130	90	100	90	120	90	90	80
	5/16	190	140	140	130	160	110	120	110	150	110	110	100	150	100	110	100	150	100	110	90
	3/8	190	130	140	120	170	110	120	100	160	100	110	100	160	100	110	90	150	100	110	90
3/4	1/4	180	140	140	130	150	110	120	110	140	100	110	100	140	100	110	90	130	90	100	90
	5/16	210	150	160	140	180	120	130	120	170	110	120	100	160	110	120	100	160	100	110	100
	3/8	210	140	160	130	180	120	130	110	170	110	120	100	170	110	120	100	160	100	110	90
1	1/4	180	140	140	140	160	120	120	120	150	120	120	110	150	110	110	110	150	110	110	100
	5/16	230	170	170	160	210	140	150	130	190	130	140	120	190	120	140	120	180	120	130	110
	3/8	230	160	170	160	210	130	150	120	200	120	140	110	190	120	140	110	180	110	130	100
1-1/4	1/4	180	140	140	140	160	120	120	120	150	120	120	110	150	110	110	110	150	110	110	100
	5/16	230	170	170	160	210	150	150	140	200	140	140	130	200	140	140	130	190	130	140	120
	3/8	230	170	170	160	210	150	150	140	200	140	140	130	200	130	140	120	190	120	140	120
1-1/2	1/4	180	140	140	140	160	120	120	120	150	120	120	110	150	110	110	110	150	110	110	100
	5/16	230	170	170	160	210	150	150	140	200	140	140	130	200	140	140	130	190	140	140	130
	3/8	230	170	170	160	210	150	150	140	200	140	140	130	200	140	140	130	190	140	140	120
	7/16	360	260	260	240	320	220	230	200	310	200	210	180	310	190	210	180	300	180	200	160
	1/2	460	310	320	280	410	250	290	230	390	220	270	200	390	220	260	200	370	210	250	190
	5/8	700	410	500	370	600	340	420	310	560	310	380	280	550	310	380	270	530	290	360	260
	3/4	950	550	660	490	830	470	560	410	770	440	510	380	760	430	510	370	730	400	480	360
	7/8	1240	720	830	630	1080	560	710	540	1020	490	660	490	1010	470	650	470	970	430	610	430
	1	1550	800	1010	780	1360	600	870	600	1290	530	810	530	1280	500	790	500	1230	470	760	470
1-3/4	1/4	180	140	140	140	160	120	120	120	150	120	120	110	150	110	110	110	150	110	110	100
	5/16	230	170	170	160	210	150	150	140	200	140	140	130	200	140	140	130	190	140	140	130
	3/8	230	170	170	160	210	150	150	140	200	140	140	130	200	140	140	130	190	140	140	120
	7/16	360	260	260	240	320	230	230	210	310	210	210	190	310	210	210	190	300	200	200	180
	1/2	460	320	320	290	410	270	290	250	390	240	270	220	390	240	260	220	380	220	250	200
	5/8	740	440	500	400	660	360	440	320	610	330	420	290	600	320	410	290	570	300	390	270
	3/4	1030	580	720	520	890	480	600	430	830	450	550	390	820	440	540	380	780	420	510	360
	7/8	1320	740	890	650	1150	630	750	550	1070	570	700	510	1060	550	680	490	1010	500	650	470
	1	1630	910	1070	790	1420	700	910	670	1340	610	850	610	1320	590	830	590	1270	550	790	550
2-1/2	1/4	180	140	140	140	160	120	120	120	150	120	120	110	150	110	110	110	150	110	110	100
	5/16	230	170	170	160	210	150	150	140	200	140	140	130	200	140	140	130	190	140	140	130
	3/8	230	170	170	160	210	150	150	140	200	140	140	130	200	140	140	130	190	140	140	120
	7/16	360	260	260	240	320	230	230	210	310	210	210	190	310	210	210	190	300	200	200	180
	1/2	460	320	320	290	410	290	290	250	390	270	270	240	390	260	260	230	380	250	250	220
	5/8	740	500	500	450	670	430	440	390	640	390	420	350	630	380	410	340	610	360	390	320
	3/4	1110	680	740	610	1010	550	650	490	960	500	610	450	950	490	600	430	920	460	580	410
	7/8	1550	830	1000	740	1370	690	880	600	1280	630	830	550	1260	620	810	530	1190	580	770	500
	1	1940	980	1270	860	1660	830	1080	720	1550	770	990	660	1520	750	970	640	1450	720	920	620
3-1/2	1/4	180	140	140	140	160	120	120	120	150	120	120	110	150	110	110	110	150	110	110	100
	5/16	230	170	170	160	210	150	150	140	200	140	140	130	200	140	140	130	190	140	140	130
	3/8	230	170	170	160	210	150	150	140	200	140	140	130	200	140	140	130	190	140	140	120
	7/16	360	260	260	240	320	230	230	210	310	210	210	190	310	210	210	190	300	200	200	180
	1/2	460	320	320	290	410	290	290	250	390	270	270	240	390	260	260	230	380	250	250	220
	5/8	740	500	500	450	670	440	440	390	640	420	420	360	630	410	410	360	610	390	390	340
	3/4	1110	740	740	650	1010	650	650	560	960	600	610	520	950	580	600	510	920	550	580	490
	7/8	1550	990	1000	860	1400	800	880	710	1340	720	830	640	1320	700	810	620	1280	660	780	570
	1	2020	1140	1270	1010	1830	930	1120	810	1740	850	1060	740	1730	830	1040	720	1670	790	1000	680

1. Tabulated lateral design values (Z) shall be multiplied by all applicable adjustment factors (see Table 10.3.1).
2. Tabulated lateral design values (Z) are for "reduced diameter body" lag screws (see Appendix L) inserted in side grain with screw axis perpendicular to wood fibers; minimum screw penetration, p, into the main member equal to 8D; screw bending yield strengths (F_{yb}): F_{yb} = 70,000 psi for D = 1/4"; F_{yb} = 60,000 psi for D = 5/16"; F_{yb} = 45,000 psi for D ≥ 3/8"
3. When 4D ≤ p < 8D, tabulated lateral design values (Z) shall be multiplied by p/8D.

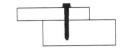

Table 11J (Cont.)   LAG SCREWS: Reference Lateral Design Values (Z) for Single Shear (two member) Connections[1,2,3]

for sawn lumber or SCL with both members of identical specific gravity

Side Member Thickness	Lag Screw Diameter	G=0.43 Hem-Fir				G=0.42 Spruce-Pine-Fir				G=0.37 Redwood (open grain)				G=0.36 Eastern Softwoods Spruce-Pine-Fir(S) Western Cedars Western Woods				G=0.35 Northern Species			
t_s	D	$Z_{\parallel}$	$Z_{s\perp}$	$Z_{m\perp}$	$Z_{\perp}$	$Z_{\parallel}$	$Z_{s\perp}$	$Z_{m\perp}$	$Z_{\perp}$	$Z_{\parallel}$	$Z_{s\perp}$	$Z_{m\perp}$	$Z_{\perp}$	$Z_{\parallel}$	$Z_{s\perp}$	$Z_{m\perp}$	$Z_{\perp}$	$Z_{\parallel}$	$Z_{s\perp}$	$Z_{m\perp}$	$Z_{\perp}$
in.	in.	lbs.	lbs.	lbs.	lbs.	lbs.	lbs.	lbs.	lbs.	lbs.	lbs.	lbs.	lbs.	lbs.	lbs.	lbs.	lbs.	lbs.	lbs.	lbs.	lbs.
1/2	1/4	110	80	80	70	110	80	80	70	100	70	70	60	100	70	70	60	90	70	70	60
	5/16	130	90	100	80	130	90	90	80	120	80	90	80	120	80	90	70	120	80	80	70
	3/8	140	80	100	80	130	80	90	80	120	60	90	60	120	60	80	60	120	60	80	60
5/8	1/4	120	80	90	80	110	80	90	70	110	70	80	70	100	70	80	60	100	70	70	60
	5/16	140	90	100	90	140	90	100	90	130	80	90	80	130	80	90	80	120	80	90	70
	3/8	140	90	100	80	140	90	100	80	130	80	90	70	130	70	90	70	120	70	90	70
3/4	1/4	130	90	100	80	120	80	90	80	110	80	90	70	110	70	80	70	110	70	80	70
	5/16	150	100	110	90	150	100	110	90	130	90	100	80	130	90	90	80	130	80	90	70
	3/8	150	100	110	90	150	90	110	90	140	90	100	80	130	80	90	70	130	80	90	70
1	1/4	140	100	110	90	140	100	100	90	130	90	100	80	130	80	90	80	130	80	90	70
	5/16	170	110	130	100	170	110	120	100	150	90	110	90	150	90	110	80	150	90	100	80
	3/8	170	100	120	100	170	100	120	90	150	90	110	80	150	90	110	80	150	90	100	80
1-1/4	1/4	140	110	110	100	140	100	100	100	130	100	100	90	130	90	90	90	130	90	90	80
	5/16	180	120	130	110	180	120	130	110	170	100	120	100	170	100	120	90	160	100	110	90
	3/8	190	120	130	110	180	110	130	100	170	100	120	90	170	100	120	90	170	90	110	80
1-1/2	1/4	140	110	110	100	140	100	100	100	130	100	100	90	130	90	90	90	130	90	90	80
	5/16	180	130	130	120	180	130	130	120	170	110	120	110	170	110	120	100	160	110	110	100
	3/8	190	130	130	120	180	130	130	110	170	110	120	100	170	110	120	100	170	100	110	90
	7/16	290	170	190	150	280	160	190	150	260	140	180	130	260	140	170	130	250	140	170	120
	1/2	350	190	240	180	350	190	240	170	310	170	210	150	310	160	210	150	300	160	200	140
	5/8	500	280	340	240	490	270	330	240	450	250	300	210	440	240	290	210	430	240	280	200
	3/4	700	360	450	330	690	350	440	330	630	290	400	290	620	280	390	280	610	270	380	270
	7/8	930	390	580	390	910	380	570	380	850	320	520	320	840	310	510	310	820	290	490	290
	1	1180	420	720	420	1160	410	710	410	1080	340	640	340	1070	330	630	330	1050	320	620	320
1-3/4	1/4	140	110	110	100	140	100	100	100	130	100	100	90	130	90	90	90	130	90	90	80
	5/16	180	130	130	120	180	130	130	120	170	120	120	110	170	120	120	110	160	110	110	100
	3/8	190	130	130	120	180	130	130	110	170	120	120	100	170	120	120	100	170	110	110	100
	7/16	290	180	190	160	280	180	190	160	270	160	180	140	260	150	170	140	260	140	170	130
	1/2	360	210	240	190	360	200	240	180	340	180	220	160	340	170	220	150	330	170	210	150
	5/8	540	290	360	250	530	280	360	250	480	250	320	220	480	250	310	210	460	240	300	210
	3/4	740	400	480	340	730	390	470	340	670	330	420	300	660	320	420	300	640	310	410	290
	7/8	970	450	610	440	950	440	600	440	880	370	540	370	870	360	530	360	850	330	520	330
	1	1210	490	750	490	1200	480	740	480	1110	400	670	400	1090	380	650	380	1070	370	640	370
2-1/2	1/4	140	110	110	100	140	100	100	100	130	100	100	90	130	90	90	90	130	90	90	80
	5/16	180	130	130	120	180	130	130	120	170	120	120	110	170	120	120	110	160	110	110	100
	3/8	190	130	130	120	180	130	130	110	170	120	120	100	170	120	120	100	170	110	110	100
	7/16	290	190	190	170	280	190	190	170	270	180	180	150	260	170	170	150	260	170	170	150
	1/2	360	240	240	210	360	240	240	210	340	220	220	190	340	210	220	190	330	200	210	180
	5/8	590	330	380	290	580	320	370	290	550	290	340	250	540	280	340	240	530	270	330	240
	3/4	890	430	550	380	880	420	540	370	800	380	500	320	780	370	490	320	760	360	480	310
	7/8	1130	550	730	470	1110	540	710	460	1010	490	640	420	990	480	620	410	970	470	600	390
	1	1380	680	870	580	1360	670	850	570	1240	570	760	510	1220	550	750	500	1190	530	730	490
3-1/2	1/4	140	110	110	100	140	100	100	100	130	100	100	90	130	90	90	90	130	90	90	80
	5/16	180	130	130	120	180	130	130	120	170	120	120	110	170	120	120	110	160	110	110	100
	3/8	190	130	130	120	180	130	130	110	170	120	120	100	170	120	120	100	170	110	110	100
	7/16	290	190	190	170	280	190	190	170	270	180	180	150	260	170	170	150	260	170	170	150
	1/2	360	240	240	210	360	240	240	210	340	220	220	190	340	220	220	190	330	210	210	180
	5/8	590	380	380	320	580	370	370	320	550	340	340	290	540	330	340	280	530	320	330	280
	3/4	890	500	550	440	880	490	540	430	830	430	500	370	820	420	490	370	800	410	480	360
	7/8	1240	610	750	530	1220	600	740	520	1150	530	680	460	1140	520	670	450	1110	500	650	430
	1	1610	740	950	630	1600	720	940	620	1480	650	860	550	1450	630	850	540	1410	620	830	520

1. Tabulated lateral design values (Z) shall be multiplied by all applicable adjustment factors (see Table 10.3.1).
2. Tabulated lateral design values (Z) are for "reduced diameter body" lag screws (see Appendix L) inserted in side grain with screw axis perpendicular to wood fibers; minimum screw penetration, p, into the main member equal to 8D; screw bending yield strengths (F_{yb}): F_{yb} = 70,000 psi for D = 1/4"; F_{yb} = 60,000 psi for D = 5/16"; F_{yb} = 45,000 psi for D ≥ 3/8"
3. When 4D ≤ p < 8D, tabulated lateral design values (Z) shall be multiplied by p/8D.

LAG SCREWS

DOWEL-TYPE FASTENERS

11

LAG SCREWS

Table 11K   LAG SCREWS: Reference Lateral Design Values (Z) for Single Shear (two member) Connections[1,2,3]

with 1/4" ASTM A 36 steel side plate, or ASTM A 653, Grade 33 steel side plate (for $t_s < 1/4$")

Side Member Thickness t_s in.	Lag Screw Diameter D in.	G=0.67 Red Oak $Z_\parallel$ lbs.	$Z_\perp$ lbs.	G=0.55 Mixed Maple Southern Pine $Z_\parallel$ lbs.	$Z_\perp$ lbs.	G=0.5 Douglas Fir-Larch $Z_\parallel$ lbs.	$Z_\perp$ lbs.	G=0.49 Douglas Fir-Larch (N) $Z_\parallel$ lbs.	$Z_\perp$ lbs.	G=0.46 Douglas Fir(S) Hem-Fir(N) $Z_\parallel$ lbs.	$Z_\perp$ lbs.	G=0.43 Hem-Fir $Z_\parallel$ lbs.	$Z_\perp$ lbs.	G=0.42 Spruce-Pine-Fir $Z_\parallel$ lbs.	$Z_\perp$ lbs.	G=0.37 Redwood (open grain) $Z_\parallel$ lbs.	$Z_\perp$ lbs.	G=0.36 Eastern Softwoods Spruce-Pine-Fir(S) Western Cedars Western Woods $Z_\parallel$ lbs.	$Z_\perp$ lbs.	G=0.35 Northern Species $Z_\parallel$ lbs.	$Z_\perp$ lbs.
0.075 (14 gage)	1/4	170	130	160	120	150	110	150	110	150	100	140	100	140	100	130	90	130	90	130	90
	5/16	220	160	200	140	190	130	190	130	190	130	180	120	180	120	170	110	170	110	160	100
	3/8	220	160	200	140	200	130	190	130	190	120	180	120	180	120	170	110	170	100	170	100
0.105 (12 gage)	1/4	180	140	170	130	160	120	160	120	160	110	150	110	150	110	140	100	140	100	140	90
	5/16	230	170	210	150	200	140	200	140	190	130	190	130	190	120	180	110	170	110	170	110
	3/8	230	160	210	140	200	140	200	130	200	130	190	120	190	120	180	110	180	110	170	110
0.120 (11 gage)	1/4	190	150	180	130	170	120	170	120	160	120	160	110	160	110	150	100	150	100	140	100
	5/16	230	170	210	150	210	140	200	140	200	140	190	130	190	130	180	120	180	120	180	110
	3/8	240	170	220	150	210	140	210	140	200	130	200	130	190	120	180	120	180	110	180	110
0.134 (10 gage)	1/4	200	150	180	140	180	130	170	130	170	120	160	120	160	110	150	110	150	100	150	100
	5/16	240	180	220	160	210	150	210	140	200	140	200	130	200	130	190	120	180	120	180	120
	3/8	240	170	220	150	220	140	210	140	210	140	200	130	200	130	190	120	190	120	180	110
0.179 (7 gage)	1/4	220	170	210	150	200	150	200	140	190	140	190	130	190	130	180	120	170	120	170	120
	5/16	260	190	240	170	230	160	230	160	230	150	220	150	220	150	210	130	200	130	200	130
	3/8	270	190	250	170	240	160	240	160	230	150	220	140	220	140	210	130	210	130	200	130
0.239 (3 gage)	1/4	240	180	220	160	210	150	210	150	200	140	190	140	190	130	180	120	180	120	180	120
	5/16	300	220	280	190	270	180	260	180	260	170	250	160	250	160	230	150	230	150	230	140
	3/8	310	220	280	190	270	180	270	180	260	170	250	160	250	160	240	140	230	140	230	140
	7/16	420	290	390	260	380	240	370	240	360	230	350	220	350	220	330	200	330	200	320	190
	1/2	510	340	470	300	460	290	450	280	440	270	430	260	420	260	400	240	400	230	390	230
	5/8	770	490	710	430	680	400	680	400	660	380	640	370	630	360	600	330	590	330	580	320
	3/4	1110	670	1020	590	980	560	970	550	950	530	920	500	910	500	860	450	850	450	840	440
	7/8	1510	880	1390	780	1330	730	1320	710	1280	690	1250	650	1230	650	1170	590	1160	590	1140	570
	1	1940	1100	1780	960	1710	910	1700	890	1650	860	1600	820	1590	810	1500	740	1480	730	1460	710
1/4	1/4	240	180	220	160	210	150	210	150	200	140	200	140	190	130	180	120	180	120	180	120
	5/16	310	220	280	200	270	180	270	180	260	170	250	170	250	160	230	150	230	150	230	140
	3/8	320	220	290	190	280	180	270	180	270	170	260	160	250	160	240	150	240	140	230	140
	7/16	480	320	440	280	420	270	420	260	410	250	390	240	390	230	370	220	360	210	360	210
	1/2	580	390	540	340	520	320	510	320	500	310	480	290	480	290	460	270	450	260	440	260
	5/8	850	530	780	470	750	440	740	440	720	420	700	400	690	400	660	370	650	360	640	350
	3/4	1200	730	1100	640	1060	600	1050	590	1020	570	990	540	980	530	930	490	920	480	900	470
	7/8	1600	930	1470	820	1410	770	1400	750	1360	720	1320	690	1310	680	1240	630	1220	620	1200	600
	1	2040	1150	1870	1000	1800	950	1780	930	1730	900	1680	850	1660	840	1570	770	1550	760	1530	740

1. Tabulated lateral design values (Z) shall be multiplied by all applicable adjustment factors (see Table 10.3.1).
2. Tabulated lateral design values (Z) are for "reduced body diameter" lag screws (see Appendix L) inserted in side grain with screw axis perpendicular to wood fibers; minimum screw penetration, p, into the main member equal to 8D; dowel bearing strengths (F_e) of 61,850 psi for ASTM A 653, Grade 33 steel and 87,000 psi for ASTM A 36 steel and screw bending yield strengths (F_{yb}): F_{yb} = 70,000 psi for D = 1/4"; F_{yb} = 60,000 psi for D = 5/16"; F_{yb} = 45,000 psi for D ≥ 3/8"
3. When 4D ≤ p < 8D, tabulated lateral design values (Z) shall be multiplied by p/8D.

Table 11L WOOD SCREWS: Reference Lateral Design Values (Z) for Single Shear(two member) Connections[1,2,3]

for sawn lumber or SCL with both members of identical specific gravity

Side Member Thickness t_s (in.)	Wood Screw Diameter D (in.)	Wood Screw Number	G=0.67 Red Oak (lbs.)	G=0.55 Mixed Maple Southern Pine (lbs.)	G=0.5 Douglas Fir-Larch (lbs.)	G=0.49 Douglas Fir-Larch(N) (lbs.)	G=0.46 Douglas Fir(S) Hem-Fir(N) (lbs.)	G=0.43 Hem-Fir (lbs.)	G=0.42 Spruce-Pine-Fir (lbs.)	G=0.37 Redwood (open grain) (lbs.)	G=0.36 Eastern Softwoods Spruce-Pine-Fir(S) Western Cedars Western Woods (lbs.)	G=0.35 Northern Species (lbs.)
1/2	0.138	6	88	67	59	57	53	49	47	41	40	38
	0.151	7	96	74	65	63	59	54	52	45	44	42
	0.164	8	107	82	73	71	66	61	59	51	50	48
	0.177	9	121	94	83	81	76	70	68	59	58	56
	0.190	10	130	101	90	87	82	75	73	64	63	60
	0.216	12	156	123	110	107	100	93	91	79	78	75
	0.242	14	168	133	120	117	110	102	99	87	86	83
5/8	0.138	6	94	76	66	64	59	53	52	44	43	41
	0.151	7	104	83	72	70	64	58	56	48	47	45
	0.164	8	120	92	80	77	72	65	63	54	53	51
	0.177	9	136	103	91	88	81	74	72	62	61	58
	0.190	10	146	111	97	94	88	80	78	67	65	63
	0.216	12	173	133	117	114	106	97	95	82	80	77
	0.242	14	184	142	126	123	115	106	103	89	87	84
3/4	0.138	6	94	79	72	71	65	58	57	47	46	44
	0.151	7	104	87	80	77	71	64	62	52	50	48
	0.164	8	120	101	88	85	78	71	69	58	56	54
	0.177	9	142	114	99	96	88	80	78	66	64	61
	0.190	10	153	122	107	103	95	86	83	71	69	66
	0.216	12	192	144	126	122	113	103	100	86	84	80
	0.242	14	203	154	135	131	122	111	108	93	91	87
1	0.138	6	94	79	72	71	67	63	61	55	54	51
	0.151	7	104	87	80	78	74	69	68	60	59	56
	0.164	8	120	101	92	90	85	80	78	67	65	62
	0.177	9	142	118	108	106	100	94	90	75	73	70
	0.190	10	153	128	117	114	108	101	97	81	78	75
	0.216	12	193	161	147	143	131	118	114	96	93	89
	0.242	14	213	178	157	152	139	126	122	102	100	95
1-1/4	0.138	6	94	79	72	71	67	63	61	55	54	52
	0.151	7	104	87	80	78	74	69	68	60	59	57
	0.164	8	120	101	92	90	85	80	78	70	68	66
	0.177	9	142	118	108	106	100	94	92	82	80	78
	0.190	10	153	128	117	114	108	101	99	88	87	84
	0.216	12	193	161	147	144	137	128	125	108	105	100
	0.242	14	213	178	163	159	151	141	138	115	111	106
1-1/2	0.138	6	94	79	72	71	67	63	61	55	54	52
	0.151	7	104	87	80	78	74	69	68	60	59	57
	0.164	8	120	101	92	90	85	80	78	70	68	66
	0.177	9	142	118	108	106	100	94	92	82	80	78
	0.190	10	153	128	117	114	108	101	99	88	87	84
	0.216	12	193	161	147	144	137	128	125	111	109	106
	0.242	14	213	178	163	159	151	141	138	123	120	117
1-3/4	0.138	6	94	79	72	71	67	63	61	55	54	52
	0.151	7	104	87	80	78	74	69	68	60	59	57
	0.164	8	120	101	92	90	85	80	78	70	68	66
	0.177	9	142	118	108	106	100	94	92	82	80	78
	0.190	10	153	128	117	114	108	101	99	88	87	84
	0.216	12	193	161	147	144	137	128	125	111	109	106
	0.242	14	213	178	163	159	151	141	138	123	120	117

1. Tabulated lateral design values (Z) shall be multiplied by all applicable adjustment factors (see Table 10.3.1).
2. Tabulated lateral design values (Z) are for rolled thread wood screws (see Appendix L) inserted in side grain with nail axis perpendicular to wood fibers; minimum screw penetration, p, into the main member equal to 10D; and screw bending yield strengths (F_{yb}): F_{yb} = 100,000 psi for 0.099" ≤ D ≤ 0.142"; F_{yb} = 90,000 psi for 0.142" < D ≤ 0.177"; F_{yb} = 80,000 psi for 0.177" < D ≤ 0.236"; F_{yb} = 70,000 psi for 0.236" < D ≤ 0.273"
3. When 6D ≤ p < 10D, tabulated lateral design values (Z) shall be multiplied by p/10D.

WOOD SCREWS

DOWEL-TYPE FASTENERS

11

Table 11M WOOD SCREWS: Reference Lateral Design Values (Z) for Single Shear (two member) Connections[1,2,3]

with ASTM A 653, Grade 33 steel side plate

Side Member Thickness t_s (in.)	Wood Screw Diameter D (in.)	Wood Screw Number	G=0.67 Red Oak (lbs.)	G=0.55 Mixed Maple Southern Pine (lbs.)	G=0.5 Douglas Fir-Larch (lbs.)	G=0.49 Douglas Fir-Larch(N) (lbs.)	G=0.46 Douglas Fir(S) Hem-Fir(N) (lbs.)	G=0.43 Hem-Fir (lbs.)	G=0.42 Spruce-Pine-Fir (lbs.)	G=0.37 Redwood (open grain) (lbs.)	G=0.36 Eastern Softwoods Spruce-Pine-Fir(S) Western Cedars Western Woods (lbs.)	G=0.35 Northern Species (lbs.)
0.036 (20 gage)	0.138	6	89	76	70	69	66	62	60	54	53	52
	0.151	7	99	84	78	76	72	68	67	60	59	57
	0.164	8	113	97	89	87	83	78	77	69	67	66
0.048 (18 gage)	0.138	6	90	77	71	70	67	63	61	55	54	53
	0.151	7	100	85	79	77	74	69	68	61	60	58
	0.164	8	114	98	90	89	84	79	78	70	69	67
0.060 (16 gage)	0.138	6	92	79	73	72	68	64	63	57	56	54
	0.151	7	101	87	81	79	75	71	70	63	61	60
	0.164	8	116	100	92	90	86	81	79	71	70	68
	0.177	9	136	116	107	105	100	94	93	83	82	79
	0.190	10	146	125	116	114	108	102	100	90	88	86
0.075 (14 gage)	0.138	6	95	82	76	75	71	67	66	59	58	57
	0.151	7	105	90	84	82	78	74	72	65	64	62
	0.164	8	119	103	95	93	89	84	82	74	73	71
	0.177	9	139	119	110	108	103	97	95	86	84	82
	0.190	10	150	128	119	117	111	105	103	92	91	88
	0.216	12	186	159	147	145	138	130	127	114	112	109
	0.242	14	204	175	162	158	151	142	139	125	123	120
0.105 (12 gage)	0.138	6	104	90	84	82	79	74	73	66	65	63
	0.151	7	114	99	92	90	86	81	80	72	71	69
	0.164	8	129	111	103	102	97	92	90	81	80	77
	0.177	9	148	128	119	116	111	105	103	93	91	89
	0.190	10	160	138	128	125	120	113	111	100	98	96
	0.216	12	196	168	156	153	146	138	135	122	120	116
	0.242	14	213	183	170	167	159	150	147	132	130	126
0.120 (11 gage)	0.138	6	110	95	89	87	83	79	77	70	68	67
	0.151	7	120	104	97	95	91	86	84	76	75	73
	0.164	8	135	117	109	107	102	96	94	85	84	82
	0.177	9	154	133	124	121	116	110	107	97	95	93
	0.190	10	166	144	133	131	125	118	116	104	103	100
	0.216	12	202	174	162	159	152	143	140	126	124	121
	0.242	14	219	189	175	172	164	155	152	137	134	131
0.134 (10 gage)	0.138	6	116	100	93	92	88	83	81	73	72	70
	0.151	7	126	110	102	100	96	91	89	80	79	77
	0.164	8	141	122	114	112	107	101	99	89	88	86
	0.177	9	160	139	129	127	121	114	112	101	100	97
	0.190	10	173	149	139	136	130	123	121	109	107	104
	0.216	12	209	180	167	164	157	148	145	131	129	126
	0.242	14	226	195	181	177	169	160	157	141	139	135
0.179 (7 gage)	0.138	6	126	107	99	97	92	86	84	76	74	72
	0.151	7	139	118	109	107	102	95	93	84	82	80
	0.164	8	160	136	126	123	117	110	108	96	95	92
	0.177	9	184	160	148	145	138	129	127	113	111	108
	0.190	10	198	172	159	156	149	140	137	122	120	117
	0.216	12	234	203	189	186	178	168	165	149	146	143
	0.242	14	251	217	202	198	190	179	176	159	156	152
0.239 (3 gage)	0.138	6	126	107	99	97	92	86	84	76	74	72
	0.151	7	139	118	109	107	102	95	93	84	82	80
	0.164	8	160	136	126	123	117	110	108	96	95	92
	0.177	9	188	160	148	145	138	129	127	113	111	108
	0.190	10	204	173	159	156	149	140	137	122	120	117
	0.216	12	256	218	201	197	187	176	172	154	151	147
	0.242	14	283	241	222	217	207	194	190	170	167	162

1. Tabulated lateral design values (Z) shall be multiplied by all applicable adjustment factors (see Table 10.3.1).

2. Tabulated lateral design values (Z) are for rolled thread wood screws (see Appendix L) inserted in side grain with screw axis perpendicular to wood fibers; minimum screw penetration, p, into the main member equal to 10D; dowel bearing strength (F_e) of 61,850 psi for ASTM A 653, Grade 33 steel and screw bending yield strengths (F_{yb}): F_{yb} = 100,000 psi for 0.099" ≤ D ≤ 0.142"; F_{yb} = 90,000 psi for 0.142" < D ≤ 0.177"; F_{yb} = 80,000 psi for 0.177" < D ≤ 0.236"; F_{yb} = 70,000 psi for 0.236" < D ≤ 0.273"

3. When 6D ≤ p < 10D, tabulated lateral design values (Z) shall be multiplied by p/10D.

WOOD SCREWS

Table 11N COMMON WIRE, BOX, or SINKER NAILS: Reference Lateral Design Values (Z) for Single Shear (two member) Connections[1,2,3,4]

for sawn lumber or SCL with both members of identical specific gravity

Side Member Thickness t_s in.	Nail Diameter D in.	Common Wire Nail Pennyweight	Box Nail Pennyweight	Sinker Nail Pennyweight	G=0.67 Red Oak lbs.	G=0.55 Mixed Maple Southern Pine lbs.	G=0.5 Douglas Fir-Larch lbs.	G=0.49 Douglas Fir-Larch (N) lbs.	G=0.46 Douglas Fir(S) Hem-Fir(N) lbs.	G=0.43 Hem-Fir lbs.	G=0.42 Spruce-Pine-Fir lbs.	G=0.37 Redwood (open grain) lbs.	G=0.36 Eastern Softwoods Spruce-Pine-Fir(S) Western Cedars Western Woods lbs.	G=0.35 Northern Species lbs.
3/4	0.099		6d	7d	73	61	55	54	51	48	47	39	38	36
	0.113	6d	8d	8d	94	79	72	71	65	58	57	47	46	44
	0.120			10d	107	89	80	77	71	64	62	52	50	48
	0.128		10d		121	101	87	84	78	70	68	57	56	54
	0.131	8d			127	104	90	87	80	73	70	60	58	56
	0.135		16d	12d	135	108	94	91	84	76	74	63	61	58
	0.148	10d	20d	16d	154	121	105	102	94	85	83	70	69	66
	0.162	16d	40d		183	138	121	117	108	99	96	82	80	77
	0.177			20d	200	153	134	130	121	111	107	92	90	87
	0.192	20d		30d	206	157	138	134	125	114	111	96	93	90
	0.207	30d		40d	216	166	147	143	133	122	119	103	101	97
	0.225	40d			229	178	158	154	144	132	129	112	110	106
	0.244	50d		60d	234	182	162	158	147	136	132	115	113	109
1	0.099		6d	7d	73	61	55	54	51	48	47	42	41	40
	0.113	6d[4]	8d	8d	94	79	72	71	67	63	61	55	54	51
	0.120			10d	107	89	81	80	76	71	69	60	59	56
	0.128		10d		121	101	93	91	86	80	79	66	64	61
	0.131	8d			127	106	97	95	90	84	82	68	66	63
	0.135		16d	12d	135	113	103	101	96	89	86	71	69	66
	0.148	10d	20d	16d	154	128	118	115	109	99	96	80	77	74
	0.162	16d	40d		184	154	141	137	125	113	109	91	89	85
	0.177			20d	213	178	155	150	138	125	121	102	99	95
	0.192	20d		30d	222	183	159	154	142	128	124	105	102	98
	0.207	30d		40d	243	192	167	162	149	135	131	111	109	104
	0.225	40d			268	202	177	171	159	144	140	120	117	112
	0.244	50d		60d	274	207	181	175	162	148	143	123	120	115
1-1/4	0.099		6d[4]	7d[4]	73	61	55	54	51	48	47	42	41	40
	0.113	6d[4]	8d	8d[4]	94	79	72	71	67	63	61	55	54	52
	0.120			10d	107	89	81	80	76	71	69	62	60	59
	0.128		10d		121	101	93	91	86	80	79	70	69	67
	0.131	8d[4]			127	106	97	95	90	84	82	73	72	70
	0.135		16d	12d	135	113	103	101	96	89	88	78	76	74
	0.148	10d	20d	16d	154	128	118	115	109	102	100	89	87	84
	0.162	16d	40d		184	154	141	138	131	122	120	103	100	95
	0.177			20d	213	178	163	159	151	141	136	113	110	105
	0.192	20d		30d	222	185	170	166	157	145	140	116	113	108
	0.207	30d		40d	243	203	186	182	169	152	147	123	119	114
	0.225	40d			268	224	200	193	177	160	155	130	127	121
	0.244	50d		60d	276	230	204	197	181	163	158	133	129	124
1-1/2	0.099			7d[4]	73	61	55	54	51	48	47	42	41	40
	0.113		8d[4]	8d[4]	94	79	72	71	67	63	61	55	54	52
	0.120			10d	107	89	81	80	76	71	69	62	60	59
	0.128		10d		121	101	93	91	86	80	79	70	69	67
	0.131	8d[4]			127	106	97	95	90	84	82	73	72	70
	0.135		16d	12d	135	113	103	101	96	89	88	78	76	74
	0.148	10d	20d	16d	154	128	118	115	109	102	100	89	87	84
	0.162	16d	40d		184	154	141	138	131	122	120	106	104	101
	0.177			20d	213	178	163	159	151	141	138	123	121	117
	0.192	20d		30d	222	185	170	166	157	147	144	128	126	120
	0.207	30d		40d	243	203	186	182	172	161	158	135	131	125
	0.225	40d			268	224	205	201	190	178	172	143	138	132
	0.244	50d		60d	276	230	211	206	196	181	175	146	141	135
1-3/4	0.113		8d[4]		94	79	72	71	67	63	61	55	54	52
	0.120			10d[4]	107	89	81	80	76	71	69	62	60	59
	0.128		10d[4]		121	101	93	91	86	80	79	70	69	67
	0.135		16d	12d	135	113	103	101	96	89	88	78	76	74
	0.148	10d[4]	20d	16d	154	128	118	115	109	102	100	89	87	84
	0.162	16d	40d		184	154	141	138	131	122	120	106	104	101
	0.177			20d	213	178	163	159	151	141	138	123	121	117
	0.192	20d		30d	222	185	170	166	157	147	144	128	126	122
	0.207	30d		40d	243	203	186	182	172	161	158	140	137	133
	0.225	40d			268	224	205	201	190	178	174	155	151	144
	0.244	50d		60d	276	230	211	206	196	183	179	159	154	147

1. Tabulated lateral design values (Z) shall be multiplied by all applicable adjustment factors (see Table 10.3.1).
2. Tabulated lateral design values (Z) are for common wire, box, and sinker nails (see Appendix L) inserted in side grain with nail axis perpendicular to wood fibers; minimum nail penetration, p, into the main member equal to 10D; and nail bending yield strengths (F_{yb}): F_{yb} = 100,000 psi for 0.099" ≤ D ≤ 0.142"; F_{yb} = 90,000 psi for 0.142" < D ≤ 0.177"; F_{yb} = 80,000 psi for 0.177" < D ≤ 0.236"; F_{yb} = 70,000 psi for 0.236" < D ≤ 0.273"
3. When 6D ≤ p < 10D, tabulated lateral design values (Z) shall be multiplied by p/10D.
4. Nail length is insufficient to provide 10D penetration. Tabulated lateral design values (Z) shall be adjusted per footnote 3.

NAILS

DOWEL-TYPE FASTENERS

11

Table 11P COMMON WIRE, BOX, or SINKER NAILS: Reference Lateral Design Values (Z) for Single Shear (two member) Connections[1,2,3]

with ASTM A 653, Grade 33 steel side plates

Side Member Thickness t_s (in.)	Nail Diameter D (in.)	Common Wire Nail (Pennyweight)	Box Nail	Sinker Nail	G=0.67 Red Oak (lbs.)	G=0.55 Mixed Maple Southern Pine (lbs.)	G=0.5 Douglas Fir-Larch (lbs.)	G=0.49 Douglas Fir-Larch (N) (lbs.)	G=0.46 Douglas Fir(S) Hem-Fir(N) (lbs.)	G=0.43 Hem-Fir (lbs.)	G=0.42 Spruce-Pine-Fir (lbs.)	G=0.37 Redwood (open grain) (lbs.)	G=0.36 Eastern Softwoods Spruce-Pine-Fir(S) Western Cedars Western Woods (lbs.)	G=0.35 Northern Species (lbs.)
0.036 (20 gage)	0.099	6d		7d	69	59	54	53	51	48	47	42	41	40
	0.113	6d	8d	8d	89	76	70	69	66	62	60	54	53	52
	0.120			10d	100	86	79	77	74	69	68	61	60	58
	0.128		10d		114	97	90	88	84	79	77	69	68	66
	0.131	8d			120	102	94	92	88	82	81	72	71	69
	0.135		16d	12d	127	108	100	98	93	87	86	77	75	73
	0.148	10d	20d	16d	145	123	114	111	106	100	98	87	86	83
0.048 (18 gage)	0.099	6d		7d	70	60	55	54	52	49	48	43	42	41
	0.113	6d	8d	8d	90	77	71	70	67	63	61	55	54	53
	0.120			10d	101	87	80	78	75	70	69	62	61	59
	0.128		10d		115	98	91	89	85	80	78	70	69	67
	0.131	8d			120	103	95	93	89	83	82	73	72	70
	0.135		16d	12d	128	109	101	99	94	88	87	78	76	74
	0.148	10d	20d	16d	145	124	115	112	107	101	99	88	87	84
	0.162	16d		40d	174	148	137	134	128	120	118	105	104	101
	0.177			20d	201	171	158	155	147	138	136	122	119	116
	0.192	20d		30d	209	178	164	161	153	144	141	126	124	121
	0.207	30d		40d	229	195	179	176	167	157	154	138	136	132
0.060 (16 gage)	0.099	6d		7d	72	62	57	56	54	51	50	45	44	43
	0.113	6d	8d	8d	92	79	73	72	68	64	63	57	56	54
	0.120			10d	103	88	82	80	76	72	71	63	62	61
	0.128		10d		117	100	92	91	86	81	80	72	70	68
	0.131	8d			122	104	97	95	90	85	83	75	73	71
	0.135		16d	12d	129	111	102	100	96	90	88	79	78	76
	0.148	10d	20d	16d	147	126	116	114	109	102	100	90	88	86
	0.162	16d		40d	175	150	138	135	129	121	119	107	105	102
	0.177			20d	202	172	159	156	149	140	137	123	121	117
	0.192	20d		30d	210	179	165	162	154	145	142	128	125	122
	0.207	30d		40d	229	195	180	177	168	158	155	139	137	133
	0.225	40d			253	215	199	195	185	174	171	153	150	146
	0.244	50d		60d	260	221	204	200	191	179	176	157	155	150
0.075 (14 gage)	0.099	6d		7d	75	65	60	59	56	53	52	47	46	45
	0.113	6d	8d	8d	95	82	76	75	71	67	66	59	58	57
	0.120			10d	106	91	85	83	79	75	73	66	65	63
	0.128		10d		120	103	95	93	89	84	82	74	73	71
	0.131	8d			125	107	99	97	93	88	86	77	76	74
	0.135		16d	12d	132	113	105	103	98	93	91	82	80	78
	0.148	10d	20d	16d	150	129	119	117	111	105	103	92	91	88
	0.162	16d		40d	178	152	141	138	132	124	122	109	107	104
	0.177			20d	204	175	162	158	151	142	139	125	123	120
	0.192	20d		30d	212	182	168	165	157	148	145	130	128	124
	0.207	30d		40d	231	198	183	179	171	161	157	141	139	135
	0.225	40d			254	217	201	197	187	176	173	155	152	148
	0.244	50d		60d	261	223	206	202	193	181	178	159	156	152
0.105 (12 gage)	0.099	6d		7d	84	73	68	67	64	60	59	53	53	51
	0.113	6d	8d	8d	104	90	84	82	79	74	73	66	65	63
	0.120			10d	115	100	93	91	87	82	80	73	71	69
	0.128		10d		129	111	103	101	97	91	90	81	79	77
	0.131	8d			134	116	107	105	101	95	93	84	82	80
	0.135		16d	12d	141	122	113	111	106	100	98	88	87	84
	0.148	10d	20d	16d	159	137	127	125	119	113	110	99	98	95
	0.162	16d		40d	187	161	149	146	140	132	129	116	114	111
	0.177			20d	213	183	169	166	159	149	147	132	130	126
	0.192	20d		30d	220	189	175	172	164	155	152	137	134	131
	0.207	30d		40d	238	205	190	186	177	167	164	147	145	141
	0.225	40d			260	223	207	203	193	182	179	161	158	153
	0.244	50d		60d	268	230	212	208	199	187	183	165	162	158

1. Tabulated lateral design values (Z) shall be multiplied by all applicable adjustment factors (see Table 10.3.1).
2. Tabulated lateral design values (Z) are for common wire, box, and sinker nails (see Appendix L) inserted in side grain with nail axis perpendicular to wood fibers; minimum nail penetration, p, into the main member equal to 10D; dowel bearing strength (F_e) of 61,850 psi for ASTM A 653, Grade 33 steel and nail bending yield strengths (F_{yb}): F_{yb} = 100,000 psi for $0.099" \leq D \leq 0.142"$; F_{yb} = 90,000 psi for $0.142" < D \leq 0.177"$; F_{yb} = 80,000 psi for $0.177" < D \leq 0.236"$; F_{yb} = 70,000 psi for $0.236" < D \leq 0.273"$
3. When $6D \leq p < 10D$, tabulated lateral design values (Z) shall be multiplied by p/10D.

Table 11P (Cont.) COMMON WIRE, BOX, or SINKER NAILS: Reference Lateral Design Values (Z) for Single Shear (two member) Connections[1,2,3]

with ASTM A 653, Grade 33 steel side plates

Side Member Thickness ts in.	Nail Diameter D in.	Common Wire Nail	Box Nail	Sinker Nail Pennyweight	G=0.67 Red Oak lbs.	G=0.55 Mixed Maple Southern Pine lbs.	G=0.5 Douglas Fir-Larch lbs.	G=0.49 Douglas Fir-Larch (N) lbs.	G=0.46 Douglas Fir(S) Hem-Fir(N) lbs.	G=0.43 Hem-Fir lbs.	G=0.42 Spruce-Pine-Fir lbs.	G=0.37 Redwood (open grain) lbs.	G=0.36 Eastern Softwoods Spruce-Pine-Fir(S) Western Cedars Western Woods lbs.	G=0.35 Northern Species lbs.
0.120 (11 gage)	0.099		6d	7d	90	78	72	71	68	64	63	57	56	53
	0.113	6d	8d	8d	110	95	89	87	83	79	77	70	68	66
	0.120			10d	121	105	97	96	91	86	85	76	75	73
	0.128		10d		134	116	108	106	101	96	94	85	83	81
	0.131	8d			140	121	112	110	105	99	97	88	86	84
	0.135		16d	12d	147	127	118	116	110	104	102	92	91	88
	0.148	10d	20d	16d	165	143	133	130	124	117	115	104	102	99
	0.162	16d	40d		193	166	154	152	145	137	134	121	119	115
	0.177			20d	218	188	174	171	163	154	151	136	134	130
	0.192	20d		30d	226	195	181	177	169	159	156	141	138	135
	0.207	30d		40d	244	210	194	191	182	172	168	151	149	145
	0.225	40d			265	228	211	207	198	186	183	164	161	157
	0.244	50d		60d	272	234	217	213	203	191	187	169	166	161
0.134 (10 gage)	0.099		6d	7d	95	82	76	74	71	66	65	58	56	54
	0.113	6d	8d	8d	116	100	93	92	88	83	81	73	72	69
	0.120			10d	127	110	102	100	96	91	89	80	79	76
	0.128		10d		140	122	113	111	106	100	98	89	87	85
	0.131	8d			146	126	117	115	110	104	102	92	90	88
	0.135		16d	12d	153	132	123	121	115	109	107	96	95	92
	0.148	10d	20d	16d	172	148	138	135	129	122	120	108	106	104
	0.162	16d	40d		199	172	160	157	150	142	139	125	123	120
	0.177			20d	224	194	180	176	169	159	156	141	138	135
	0.192	20d		30d	232	200	186	182	174	164	161	145	143	139
	0.207	30d		40d	249	215	199	196	187	176	173	156	153	149
	0.225	40d			270	233	216	212	202	191	187	168	165	161
	0.244	50d		60d	277	239	221	217	207	195	192	173	170	165
0.179 (7 gage)	0.099		6d	7d	97	82	76	74	71	66	65	58	56	54
	0.113	6d	8d	8d	126	107	99	97	92	86	84	76	74	70
	0.120			10d	142	121	111	109	104	97	95	85	83	79
	0.128		10d		161	137	126	124	118	111	108	97	94	90
	0.131	8d			168	144	132	130	123	116	114	102	99	94
	0.135		16d	12d	175	152	141	138	131	123	121	108	105	100
	0.148	10d	20d	16d	195	170	158	155	148	140	137	123	121	117
	0.162	16d	40d		224	194	180	177	169	160	157	142	140	136
	0.177			20d	249	215	200	197	188	178	174	157	155	151
	0.192	20d		30d	256	222	206	203	194	183	179	162	159	155
	0.207	30d		40d	272	236	219	215	205	194	190	172	169	164
	0.225	40d			292	252	234	230	220	207	203	184	180	176
	0.244	50d		60d	299	258	240	235	225	212	208	188	185	180
0.239 (3 gage)	0.099		6d	7d	97	82	76	74	71	66	65	58	56	54
	0.113	6d	8d	8d	126	107	99	97	92	86	84	76	74	70
	0.120			10d	142	121	111	109	104	97	95	85	83	79
	0.128		10d		161	137	126	124	118	111	108	97	94	90
	0.131	8d			169	144	132	130	123	116	114	102	99	94
	0.135		16d	12d	180	153	141	138	131	123	121	108	105	100
	0.148	10d	20d	16d	205	174	160	157	149	140	137	123	121	117
	0.162	16d	40d		245	209	192	188	179	168	165	147	145	140
	0.177			20d	284	241	222	218	207	195	191	170	167	162
	0.192	20d		30d	295	251	231	227	216	202	198	177	174	169
	0.207	30d		40d	310	270	251	246	236	222	217	194	191	185
	0.225	40d			328	285	265	260	249	235	231	209	205	200
	0.244	50d		60d	336	291	271	266	254	240	236	213	210	204

1. Tabulated lateral design values (Z) shall be multiplied by all applicable adjustment factors (see Table 10.3.1).
2. Tabulated lateral design values (Z) are for common wire, box, and sinker nails (see Appendix L) inserted in side grain with nail axis perpendicular to wood fibers; minimum nail penetration, p, into the main member equal to 10D; dowel bearing strength (F_e) of 61,850 psi for ASTM A 653, Grade 33 steel and nail bending yield strengths (F_{yb}): F_{yb} = 100,000 psi for 0.099" ≤ D ≤ 0.142"; F_{yb} = 90,000 psi for 0.142" < D ≤ 0.177"; F_{yb} = 80,000 psi for 0.177" < D ≤ 0.236"; F_{yb} = 70,000 psi for 0.236" < D ≤ 0.273"
3. When 6D ≤ p < 10D, tabulated lateral design values (Z) shall be multiplied by p/10D.

NAILS

DOWEL-TYPE FASTENERS

11

Table 11Q COMMON WIRE, BOX, or SINKER NAILS: Reference Lateral Design Values (Z) for Single Shear (two member) Connections[1,2,3,4]

with wood structural panel side members with an effective G = 0.50

NAILS

Side Member Thickness t_s (in.)	Nail Diameter D (in.)	Common Wire Nail (Pennyweight)	Box Nail (Pennyweight)	Sinker Nail (Pennyweight)	G=0.67 Red Oak (lbs.)	G=0.55 Mixed Maple Southern Pine (lbs.)	G=0.5 Douglas Fir-Larch (lbs.)	G=0.49 Douglas Fir-Larch (N) (lbs.)	G=0.46 Douglas Fir(S) Hem-Fir(N) (lbs.)	G=0.43 Hem-Fir (lbs.)	G=0.42 Spruce-Pine-Fir (lbs.)	G=0.37 Redwood (open grain) (lbs.)	G=0.36 Eastern Softwoods Spruce-Pine-Fir(S) Western Cedars Western Woods (lbs.)	G=0.35 Northern Species (lbs.)
3/8	0.099		6d	7d	47	45	43	43	42	40	40	38	37	37
	0.113	6d	8d	8d	60	56	54	54	52	51	50	47	47	46
	0.120			10d	67	62	60	60	58	56	56	52	52	51
	0.128		10d		75	70	68	67	65	63	63	59	58	57
	0.131	8d			78	73	71	70	68	66	65	61	61	60
	0.135		16d	12d	83	78	75	74	72	70	69	65	64	63
	0.148	10d	20d	16d	94	88	85	84	82	79	78	73	72	71
7/16	0.099		6d	7d	50	47	45	45	44	43	42	40	40	39
	0.113	6d	8d	8d	62	58	56	56	55	53	52	49	49	48
	0.120			10d	69	65	63	62	60	59	58	55	54	53
	0.128		10d		77	72	70	69	66	66	65	61	60	59
	0.131	8d			80	75	73	72	70	68	67	63	63	62
	0.135		16d	12d	85	80	77	76	74	72	71	67	66	65
	0.148	10d	20d	16d	96	90	87	86	84	81	80	76	75	73
	0.162	16d	40d		114	106	102	101	99	96	95	89	88	86
15/32	0.099		6d	7d	51	48	47	46	45	44	44	41	41	40
	0.113	6d	8d	8d	64	60	58	57	56	54	54	51	50	49
	0.120			10d	70	66	64	63	62	60	59	56	55	54
	0.128		10d		78	74	71	71	69	67	66	62	62	61
	0.131	8d			82	77	74	73	72	70	69	65	64	63
	0.135		16d	12d	86	81	78	77	76	73	72	68	67	66
	0.148	10d	20d	16d	97	91	88	87	85	83	82	77	76	75
	0.162	16d	40d		115	108	104	103	100	97	96	90	89	88
19/32	0.099		6d	7d	58	55	53	53	51	50	50	47	46	46
	0.113	6d	8d	8d	70	66	64	64	62	61	60	57	56	55
	0.120			10d	77	73	70	70	68	66	66	62	61	60
	0.128		10d		85	80	78	77	75	73	72	68	68	67
	0.131	8d			88	83	80	80	78	76	75	71	70	69
	0.135		16d	12d	93	87	84	84	82	79	79	74	73	72
	0.148	10d	20d	16d	104	98	95	94	92	89	88	83	82	81
	0.162	16d	40d		121	114	110	109	107	103	102	96	95	94
	0.177			20d	137	128	124	123	120	116	115	108	107	105
	0.192	20d		30d	142	133	128	127	124	120	119	112	111	109
23/32	0.099		6d	7d	62	58	55	55	53	51	51	47	47	46
	0.113	6d	8d	8d	78	74	72	71	69	67	66	62	61	60
	0.120			10d	85	80	78	77	76	73	73	69	68	67
	0.128		10d		93	88	85	85	83	80	80	75	75	74
	0.131	8d			96	91	88	87	86	83	82	78	77	76
	0.135		16d	12d	101	95	92	91	89	87	86	81	81	79
	0.148	10d	20d	16d	113	106	103	102	100	97	96	91	90	89
	0.162	16d	40d		130	122	118	117	115	111	110	104	103	102
	0.177			20d	145	137	132	131	128	124	123	116	115	113
	0.192	20d		30d	150	141	136	135	132	128	127	120	118	116
1	0.099[5]		6d	7d	62	58	55	55	53	51	51	47	47	46
	0.113[5]	6d[4]	8d	8d	81	75	72	71	69	67	66	62	61	60
	0.120[5]			10d	92	85	81	81	78	76	75	69	69	67
	0.128		10d		104	97	93	92	89	86	85	79	78	77
	0.131	8d			109	101	97	96	93	90	89	83	82	80
	0.135		16d	12d	116	108	103	102	99	96	94	88	87	85
	0.148	10d	20d	16d	132	123	118	116	113	109	108	100	99	97
	0.162	16d	40d		154	146	141	139	135	131	129	120	119	116
	0.177			20d	169	160	155	154	151	146	145	137	136	134
	0.192	20d		30d	174	164	159	158	155	150	149	141	140	138
1-1/8	0.128[5]		10d		104	97	93	92	89	86	85	79	78	77
	0.131[5]	8d			109	101	97	96	93	90	89	83	82	80
	0.135[5]		16d	12d	116	108	103	102	99	96	94	88	87	85
	0.148[5]	10d	20d	16d	132	123	118	116	113	109	108	100	99	97
	0.162	16d	40d		158	147	141	139	135	131	129	120	119	116
	0.177			20d	181	170	163	161	157	151	149	139	137	135
	0.192	20d		30d	186	176	170	168	163	157	155	145	143	140
1-1/4	0.148	10d	20d	16d	132	123	118	116	113	109	108	100	99	97
	0.162	16d	40d		158	147	141	139	135	131	129	120	119	116
	0.177			20d	183	170	163	161	157	151	149	139	137	135
	0.192	20d		30d	191	177	170	168	163	157	155	145	143	140

1. Tabulated lateral design values (Z) shall be multiplied by all applicable adjustment factors (see Table 10.3.1).
2. Tabulated lateral design values (Z) are for common wire, box, and sinker nails (see Appendix L) inserted in side grain with nail axis perpendicular to wood fibers; minimum nail penetration, p, into the main member equal to 10D and nail bending yield strengths (F_{yb}): F_{yb} = 100,000 psi for 0.099" ≤ D ≤ 0.142"; F_{yb} = 90,000 psi for 0.142" < D ≤ 0.177"; F_{yb} = 80,000 psi for 0.177" < D ≤ 0.236".
3. When 6D ≤ p < 10D, tabulated lateral design values (Z) shall be multiplied by p/10D.
4. Nail length is insufficient to provide 10D penetration. Tabulated lateral design values (Z) shall be adjusted per footnote 3.
5. Tabulated lateral design values (Z) shall be permitted to apply for greater side member thickness when adjusted per footnote 3.

Table 11R COMMON WIRE, BOX, or SINKER NAILS: Reference Lateral Design Values (Z) for Single Shear (two member) Connections[1,2,3,4]

with wood structural panel side members with an effective G = 0.42

Side Member Thickness t_s (in.)	Nail Diameter D (in.)	Common Wire Nail	Box Nail	Sinker Nail	G=0.67 Red Oak (lbs.)	G=0.55 Mixed Maple Southern Pine (lbs.)	G=0.5 Douglas Fir-Larch (lbs.)	G=0.49 Douglas Fir-Larch (N) (lbs.)	G=0.46 Douglas Fir(S) Hem-Fir(N) (lbs.)	G=0.43 Hem-Fir (lbs.)	G=0.42 Spruce-Pine-Fir (lbs.)	G=0.37 Redwood (open grain) (lbs.)	G=0.36 Eastern Softwoods Spruce-Pine-Fir(S) Western Cedars Western Woods (lbs.)	G=0.35 Northern Species (lbs.)
3/8	0.099		6d	7d	41	39	37	37	36	35	35	33	33	32
	0.113	6d	8d	8d	52	49	48	47	46	45	45	42	42	41
	0.120			10d	58	55	53	53	52	50	50	47	47	46
	0.128		10d		66	62	60	60	59	57	56	53	53	52
	0.131	8d			69	65	63	63	61	59	59	56	55	54
	0.135		16d	12d	73	69	67	66	65	63	62	59	58	57
	0.148	10d	20d	16d	84	79	76	76	74	72	71	67	66	65
7/16	0.099		6d	7d	42	40	39	38	38	37	36	35	34	34
	0.113	6d	8d	8d	53	50	49	48	48	46	46	43	43	42
	0.120			10d	59	56	54	54	53	51	51	48	48	47
	0.128		10d		67	63	61	61	60	58	57	54	54	53
	0.131	8d			70	66	64	64	62	60	60	57	56	55
	0.135		16d	12d	74	70	68	67	66	64	63	60	59	58
	0.148	10d	20d	16d	84	80	77	76	75	73	72	68	67	66
	0.162	16d	40d		100	95	92	91	89	86	85	81	80	78
15/32	0.099		6d	7d	43	41	40	39	39	38	37	35	35	35
	0.113	6d	8d	8d	54	51	50	49	48	47	47	44	44	43
	0.120			10d	60	57	55	55	54	52	52	49	49	48
	0.128		10d		68	64	62	62	60	59	58	55	55	54
	0.131	8d			70	67	65	64	63	61	61	57	57	56
	0.135		16d	12d	75	71	68	68	66	65	64	61	60	59
	0.148	10d	20d	16d	85	80	78	77	75	73	72	69	68	67
	0.162	16d	40d		101	95	92	91	89	87	86	81	80	79
19/32	0.099		6d	7d	47	45	44	43	43	41	41	39	39	38
	0.113	6d	8d	8d	58	55	54	53	52	51	50	48	48	47
	0.120			10d	64	61	59	59	58	56	56	53	52	52
	0.128		10d		71	68	66	65	64	62	62	59	58	57
	0.131	8d			74	70	68	68	67	65	64	61	61	60
	0.135		16d	12d	78	74	72	71	70	68	68	64	64	63
	0.148	10d	20d	16d	88	84	81	81	79	77	76	72	72	71
	0.162	16d	40d		103	98	95	94	93	90	89	85	84	83
	0.177			20d	118	112	108	108	105	102	101	96	95	94
	0.192	20d		30d	123	116	112	112	109	106	105	100	99	97
23/32	0.099		6d	7d	52	50	48	48	47	46	46	44	43	43
	0.113	6d	8d	8d	63	60	58	58	57	56	55	53	52	52
	0.120			10d	69	66	64	64	62	61	60	58	57	56
	0.128		10d		76	73	71	70	69	67	67	63	63	62
	0.131	8d			79	75	73	73	71	70	69	66	65	64
	0.135		16d	12d	83	79	77	76	75	73	72	69	68	67
	0.148	10d	20d	16d	93	89	86	86	84	82	81	77	77	76
	0.162	16d	40d		108	103	100	99	98	95	94	90	89	87
	0.177			20d	122	116	113	112	110	107	106	101	100	98
	0.192	20d		30d	127	120	117	116	114	111	110	104	103	102
1	0.099[5]		6d	7d	56	53	51	50	49	48	47	44	44	43
	0.113[5]	6d[4]	8d	8d	73	68	66	66	64	62	61	58	57	56
	0.120[5]			10d	82	77	75	74	72	70	69	65	64	63
	0.128		10d		91	87	85	84	82	80	79	74	73	72
	0.131	8d			93	89	87	87	85	83	82	77	77	75
	0.135		16d	12d	97	93	91	90	89	87	86	82	81	80
	0.148	10d	20d	16d	109	104	101	101	99	97	96	91	91	90
	0.162	16d	40d		124	118	115	115	113	110	109	104	103	102
	0.177			20d	137	131	128	127	125	122	121	115	114	112
	0.192	20d		30d	141	135	131	131	128	125	124	118	117	116
1-1/8	0.128[5]		10d		93	88	85	84	82	80	79	74	73	72
	0.131[5]	8d			98	92	89	88	86	83	82	77	77	75
	0.135[5]		16d	12d	104	98	94	94	91	88	88	82	81	80
	0.148[5]	10d	20d	16d	117	111	108	107	104	101	100	94	93	91
	0.162	16d	40d		132	127	123	123	120	118	117	111	110	109
	0.177			20d	146	139	136	136	132	129	128	122	121	120
	0.192	20d		30d	150	143	139	138	136	133	132	126	125	123
1-1/4	0.148	10d	20d	16d	118	111	108	107	104	101	100	94	93	91
	0.162	16d	40d		141	134	129	128	125	121	120	112	111	109
	0.177			20d	155	148	144	143	141	138	136	130	129	126
	0.192	20d		30d	159	152	148	147	144	141	140	134	133	131

1. Tabulated lateral design values (Z) shall be multiplied by all applicable adjustment factors (see Table 10.3.1).
2. Tabulated lateral design values (Z) are for common wire, box, and sinker nails (see Appendix L) inserted in side grain with nail axis perpendicular to wood fibers; minimum nail penetration, p, into the main member equal to 10D and nail bending yield strengths (F_{yb}): F_{yb} = 100,000 psi for 0.099" ≤ D ≤ 0.142"; F_{yb} = 90,000 psi for 0.142" < D ≤ 0.177"; F_{yb} = 80,000 psi for 0.177" < D ≤ 0.236"
3. When 6D ≤ p < 10D, tabulated lateral design values (Z) shall be multiplied by p/10D.
4. Nail length is insufficient to provide 10D penetration. Tabulated lateral design values (Z) shall be adjusted per footnote 3.
5. Tabulated lateral design values (Z) shall be permitted to apply for greater side member thickness when adjusted per footnote 3.

SPLIT RING AND SHEAR PLATE CONNECTORS

12

12.1 General

12.1.1 Terminology

A connector unit shall be defined as one of the following:
(a) One split ring with its bolt or lag screw in single shear (see Figure 12A).
(b) Two shear plates used back to back in the contact faces of a wood-to-wood connection with their bolt or lag screw in single shear (see Figures 12B and 12C).
(c) One shear plate with its bolt or lag screw in single shear used in conjunction with a steel strap or shape in a wood-to-metal connection (see Figures 12B and 12C).

Figure 12A Split Ring Connector

Figure 12B Pressed Steel Shear Plate Connector

Figure 12C Malleable Iron Shear Plate Connector

12.1.2 Quality of Split Ring and Shear Plate Connectors

12.1.2.1 Design provisions and reference design values herein apply to split ring and shear plate connectors of the following quality:
(a) Split rings manufactured from SAE 1010 hot rolled carbon steel (Reference 34). Each ring shall form a closed true circle with the principal axis of the cross section of the ring metal parallel to the geometric axis of the ring. The ring shall fit snugly in the precut groove. This shall be accomplished with a ring, the metal section of which is beveled from the central portion toward the edges to a thickness less than at midsection, or by any other method which will accomplish equivalent performance. It shall be cut through in one place in its circumference to form a tongue and slot (see Figure 12A).
(b) Shear plate connectors:
   (1) 2-5/8" Pressed Steel Type—Pressed steel shear plates manufactured from SAE 1010 hot rolled carbon steel. Each plate shall be a true circle with a flange around the edge, extending at right angles to the face of the plate and extending from one face only, the plate portion having a central bolt hole, with an integral hub concentric to the hole or without an integral hub, and two small perforations on opposite sides of the hole and midway from the center and circumference (see Figure 12B).
   (2) 4" Malleable Iron Type—Malleable iron shear plates manufactured according to Grade 32510 of ASTM Standard A47. Each casting shall consist of a perforated round plate with a flange around the edge extending at right angles to the face of the plate and projecting from one face only, the plate portion having a central bolt hole with an integral hub extending from the same face as the flange (see Figure 12C).

12.1.2.2 Dimensions for typical split ring and shear plate connectors are provided in Appendix K. Dimensional tolerances of split ring and shear plate connectors shall not be greater than those conforming to standard practices for the machine operations involved in manufacturing the connectors.

12.1.2.3 Bolts used with split ring and shear plate connectors shall conform to 11.1.2. The bolt shall have an unreduced nominal or shank (body) diameter in accordance with ANSI/ASME Standard B18.2.1.

12.1.2.4 When lag screws are used in place of bolts, the lag screws shall conform to 11.1.3 and the shank of the lag screw shall have the same diameter as the bolt specified for the split ring or shear plate connector (see Tables 12.2A and 12.2B). The lag screw shall have an unreduced nominal or shank (body) diameter and threads in accordance with ANSI/ASME Standard B18.2.1.

12.1.3 Fabrication and Assembly

12.1.3.1 The grooves, daps, and bolt holes specified in Appendix K shall be accurately cut or bored, and shall be oriented in contacting faces. Since split ring and shear plate connectors from different manufacturers differ slightly in shape and cross section, cutter heads shall be designed to produce daps and grooves conforming accurately to the dimensions and shape of the particular split ring or shear plate connectors used.

12.1.3.2 When lag screws are used in place of bolts, the hole for the unthreaded shank shall be the same diameter as the shank. The diameter of the hole for the threaded portion of the lag screw shall be approxi-

mately 70% of the shank diameter, or as specified in 11.1.3.2.

12.1.3.3 In installation of split ring or shear plate connectors and bolts or lag screws, a nut shall be placed on each bolt, and washers, not smaller than the size specified in Appendix K, shall be placed between the outside wood member and the bolt or lag screw head and between the outside wood member and nut. When an outside member of a shear plate connection is a steel strap or shape, the washer is not required, except when a longer bolt or lag screw is used, in which case, the washer prevents the metal plate or shape from bearing on the threaded portion of the bolt or lag screw.

12.1.3.4 Reference design values for split ring and shear plate connectors are based on the assumption that the faces of the members are brought into contact when the connector units are installed, and allow for seasonal variations after the wood has reached the moisture content normal to the conditions of service. When split ring or shear plate connectors are installed in wood which is not seasoned to the moisture content normal to the conditions of service, the connections shall be tightened by turning down the nuts periodically until moisture equilibrium is reached.

12.2 Reference Design Values

12.2.1 Reference Design Values

12.2.1.1 Tables 12.2A and 12.2B contain reference design values for a single split ring or shear plate connector unit with bolt in single shear, installed in the side grain of two wood members (Table 12A) with sufficient member thicknesses, edge distances, end distances, and spacing to develop reference design values. Reference design values (P, Q) shall be multiplied by all applicable adjustment factors (see Table 10.3.1) to obtain adjusted design values (P', Q').

12.2.1.2 Adjusted design values (P', Q') for shear plate connectors shall not exceed the limiting reference design values specified in Footnote 2 of Table 12.2B. The limiting reference design values in Footnote 2 of Table 12.2B shall not be multiplied by adjustment factors in this Specification since they are based on strength of metal rather than strength of wood (see 10.2.3).

Table 12A Species Groups for Split Ring and Shear Plate Connectors

Species Group	Specific Gravity, G
A	$G \geq 0.60$
B	$0.49 \leq G < 0.60$
C	$0.42 \leq G < 0.49$
D	$G < 0.42$

12.2.2 Thickness of Wood Members

12.2.2.1 Reference design values shall not be used for split ring or shear plate connectors installed in any piece of wood of a net thickness less than the minimum specified in Tables 12.2A and 12.2B.

12.2.2.2 Reference design values for split ring or shear plate connectors installed in any piece of wood of net thickness intermediate between the minimum thickness and that required for maximum reference design value, as specified in Tables 12.2A and 12.2B, shall be obtained by linear interpolation.

Table 12.2A Split Ring Connector Unit Reference Design Values

Tabulated design values[1] apply to ONE split ring and bolt in single shear.

Split ring diameter in.	Bolt diameter in.	Number of faces of member with connectors on same bolt	Net thickness of member in.	Loaded parallel to grain (0°) Design value, P, per connector unit and bolt, lbs.				Loaded perpendicular to grain (90°) Design value, Q, per connector unit and bolt, lbs.			
				Group A species	Group B species	Group C species	Group D species	Group A species	Group B species	Group C species	Group D species
2-1/2	1/2	1	1" minimum	2630	2270	1900	1640	1900	1620	1350	1160
			1-1/2" or thicker	3160	2730	2290	1960	2280	1940	1620	1390
		2	1-1/2" minimum	2430	2100	1760	1510	1750	1500	1250	1070
			2" or thicker	3160	2730	2290	1960	2280	1940	1620	1390
4	3/4	1	1" minimum	4090	3510	2920	2520	2840	2440	2040	1760
			1-1/2"	6020	5160	4280	3710	4180	3590	2990	2580
			1-5/8" or thicker	6140	5260	4380	3790	4270	3660	3050	2630
		2	1-1/2" minimum	4110	3520	2940	2540	2980	2450	2040	1760
			2"	4950	4250	3540	3050	3440	2960	2460	2120
			2-1/2"	5830	5000	4160	3600	4050	3480	2890	2500
			3" or thicker	6140	5260	4380	3790	4270	3660	3050	2630

1. Tabulated lateral design values (P,Q) for split ring connector units shall be multiplied to all applicable adjustment factors (see Table 10.3.1).

Table 12.2B Shear Plate Connector Unit Reference Design Values

Tabulated design values[1,2,3] apply to ONE shear plate and bolt in single shear.

Shear plate diameter in.	Bolt diameter in.	Number of faces of member with connectors on same bolt	Net thickness of member in.	Loaded parallel to grain (0°) Design value, P, per connector unit and bolt, lbs.				Loaded perpendicular to grain (90°) Design value, Q, per connector unit and bolt, lbs.			
				Group A species	Group B species	Group C species	Group D species	Group A species	Group B species	Group C species	Group D species
2-5/8	3/4	1	1-1/2" minimum	3110*	2670	2220	2010	2170	1860	1550	1330
		2	1-1/2" minimum	2420	2080	1730	1500	1690	1450	1210	1040
			2"	3190*	2730	2270	1960	2220	1910	1580	1370
			2-1/2" or thicker	3330*	2860	2380	2060	2320	1990	1650	1440
4	3/4 or 7/8	1	1-1/2" minimum	4370	3750	3130	2700	3040	2620	2170	1860
			1-3/4" or thicker	5090*	4360	3640	3140	3540	3040	2530	2200
		2	1-3/4" minimum	3390	2910	2420	2090	2360	2020	1680	1410
			2"	3790	3240	2700	2330	2640	2260	1880	1630
			2-1/2"	4310	3690	3080	2660	3000	2550	2140	1850
			3"	4830*	4140	3450	2980	3360	2880	2400	2060
			3-1/2" or thicker	5030*	4320	3600	3110	3500	3000	2510	2160

1. Tabulated lateral design values (P,Q) for shear plate connector units shall be multiplied to all applicable adjustment factors (see Table 10.3.1).
2. Allowable design values for shear plate connector units shall not exceed the following:
   (a) 2-5/8" shear plate ....2900 pounds
   (b) 4" shear plate with 3/4" bolt ....4400 pounds
   (c) 4" shear plate with 7/8" bolt ....6000 pounds
   The design values in Footnote 2 shall be permitted to be increased in accordance with the American Institute of Steel Construction (AISC) Manual of Steel Construction, 9th edition, Section A5.2 "Wind and Seismic Stresses", except when design loads have already been reduced by load combination factors (see NDS 10.2.3).
3. Loads followed by an asterisk (*) exceed those permitted by Footnote 2, but are needed for determination of design values for other angles of load to grain. Footnote 2 limitations apply in all cases.

SPLIT RING AND SHEAR PLATE CONNECTORS

12

12.2.3 Penetration Depth Factor, C_d

When lag screws instead of bolts are used with split ring or shear plate connectors, reference design values shall be multiplied by the appropriate penetration depth factor, C_d, specified in Table 12.2.3. Lag screw penetration into the member receiving the point shall not be less than the minimum penetration specified in Table 12.2.3. When the actual lag screw penetration into the member receiving the point is greater than the minimum penetration, but less than the minimum penetration for $C_d = 1.0$, the penetration depth factor, C_d, shall be determined by linear interpolation. The penetration depth factor shall not exceed unity, $C_d \leq 1.0$.

12.2.4 Metal Side Plate Factor, C_{st}

When metal side members are used in place of wood side members, the reference design values parallel to grain, P, for 4" shear plate connectors shall be multiplied by the appropriate metal side plate factor specified in Table 12.2.4.

Table 12.2.4 Metal Side Plate Factors, C_{st}, for 4" Shear Plate Connectors Loaded Parallel to Grain

Species Group	C_{st}
A	1.18
B	1.11
C	1.05
D	1.00

The adjusted design values parallel to grain, P', shall not exceed the limiting reference design values given in Footnote 2 of Table 12.2B (see 12.2.1.2).

12.2.5 Load at Angle to Grain

12.2.5.1 When a load acts in the plane of the wood surface at an angle to grain other than 0° or 90°, the adjusted design value, N', for a split ring or shear plate connector unit shall be determined as follows (see Appendix J):

$$N' = \frac{P'Q'}{P' \sin^2 \theta + Q' \cos^2 \theta} \qquad (12.2\text{-}1)$$

where:

θ = angle between direction of load and direction of grain (longitudinal axis of member)

Table 12.2.3 Penetration Depth Factors, C_d, for Split Ring and Shear Plate Connectors Used with Lag Screws

	Side Member	Penetration	Penetration of Lag Screw into Main Member (number of shank diameters) Species Group (see Table 12A)				Penetration Depth Factor, C_d
			Group A	Group B	Group C	Group D	
2-1/2" Split Ring 4" Split Ring 4" Shear Plate	Wood or Metal	Minimum for $C_d = 1.0$	7	8	10	11	1.0
		Minimum for $C_d = 0.75$	3	3-1/2	4	4-1/2	0.75
2-5/8" Shear Plate	Wood	Minimum for $C_d = 1.0$	4	5	7	8	1.0
		Minimum for $C_d = 0.75$	3	3-1/2	4	4-1/2	0.75
	Metal	Minimum for $C_d = 1.0$	3	3-1/2	4	4-1/2	1.0

12.2.5.2 Adjusted design values at an angle to grain, N', for shear plate connectors shall not exceed the limiting reference design values specified in Footnote 2 of Table 12.2.B (see 12.2.1.2).

12.2.6 Split Ring and Shear Plate Connectors in End Grain

12.2.6.1 When split ring or shear plate connectors are installed in a surface that is not parallel to the general direction of the grain of the member, such as the end of a square-cut member, or the sloping surface of a member cut at an angle to its axis, or the surface of a structural glued laminated timber cut at an angle to the direction of the laminations, the following terminology shall apply:
- "Side grain surface" means a surface parallel to the general direction of the wood fibers ($\alpha = 0°$), such as the top, bottom, and sides of a straight beam.
- "Sloping surface" means a surface cut at an angle, α, other than $0°$ or $90°$ to the general direction of the wood fibers.
- "Square-cut surface" means a surface perpendicular to the general direction of the wood fibers ($\alpha = 90°$).
- "Axis of cut" defines the direction of a sloping surface relative to the general direction of the wood fibers. For a sloping cut symmetrical about one of the major axes of the member, as in Figures 12D, 12G, 12H, and 12I, the axis of cut is parallel to a major axis. For an asymmetrical sloping surface (i.e., one that slopes relative to both major axes of the member), the axis of cut is the direction of a line defining the intersection of the sloping surface with any plane that is both normal to the sloping surface and also is aligned with the general direction of the wood fibers (see Figure 12E).

α = the least angle formed between a sloping surface and the general direction of the wood fibers (i.e., the acute angle between the axis of cut and the general direction of the fibers. Sometimes called the slope of the cut. See Figures 12D through 12I).

φ = the angle between the direction of applied load and the axis of cut of a sloping surface, measured in the plane of the sloping surface (see Figure 12I).

P' = adjusted design value for a split ring or shear plate connector unit in a side grain surface, loaded parallel to grain ($\alpha = 0°$, $\varphi = 0°$).

Q' = adjusted design value for a split ring or shear plate connector unit in a side grain surface, loaded perpendicular to grain ($\alpha = 0°$, $\varphi = 90°$).

Q'_{90} = adjusted design value for a split ring or shear plate connector unit in a square-cut surface, loaded in any direction in the plane of the surface ($\alpha = 90°$).

P'_α = adjusted design value for a split ring or shear plate connector unit in a sloping surface, loaded in a direction parallel to the axis of cut ($0° < \alpha < 90°$, $\varphi = 0°$).

Q'_α = adjusted design value for a split ring or shear plate connector unit in a sloping surface, loaded in a direction perpendicular to the axis of cut ($0° < \alpha < 90°$, $\varphi = 90°$).

N'_α = adjusted design value for a split ring or shear plate connector unit in a sloping surface, when direction of load is at an angle φ from the axis of cut.

Figure 12D Axis of Cut for Symmetrical Sloping End Cut

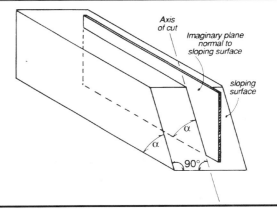

Figure 12E Axis of Cut for Asymmetrical Sloping End Cut

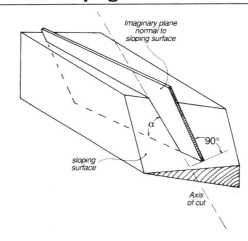

12.2.6.2 When split ring or shear plate connectors are installed in square-cut end grain or sloping surfaces, adjusted design values shall be determined as follows (see 10.2.2):

(a) Square-cut surface; loaded in any direction ($\alpha = 90°$, see Figure 12F).

$$Q_{90}' = 0.60Q' \tag{12.2-2}$$

Figure 12F   Square End Cut

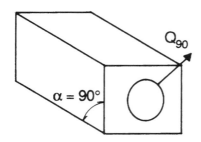

(b) Sloping surface; loaded parallel to axis of cut ($0° < \alpha < 90°$, $\varphi = 0°$, see Figure 12G).

$$P_\alpha' = \frac{P'Q_{90}'}{P' \sin^2 \alpha + Q_{90}' \cos^2 \alpha} \tag{12.2-3}$$

Figure 12G   Sloping End Cut with Load Parallel to Axis of Cut ($\varphi = 0°$)

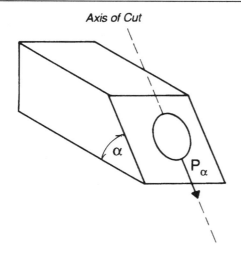

(c) Sloping surface; loaded perpendicular to axis of cut ($0° < \alpha < 90°$, $\varphi = 90°$, see Figure 12H).

$$Q_\alpha' = \frac{Q'Q_{90}'}{Q' \sin^2 \alpha + Q_{90}' \cos^2 \alpha} \tag{12.2-4}$$

Figure 12H   Sloping End Cut with Load Perpendicular to Axis of Cut ($\varphi = 90°$)

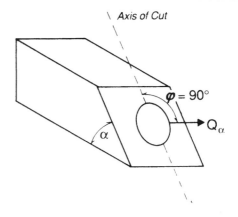

(d) Sloping surface; loaded at angle φ to axis of cut ($0° < \alpha < 90°$, $0° < \varphi < 90°$, see Figure 12I).

$$N_\alpha' = \frac{P_\alpha'Q_\alpha'}{P_\alpha' \sin^2 \varphi + Q_\alpha' \cos^2 \varphi} \tag{12.2-5}$$

Figure 12I   Sloping End Cut with Load at an Angle φ to Axis of Cut

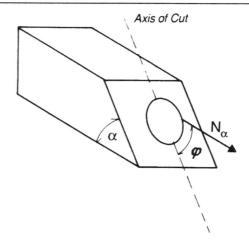

12.3 Placement of Split Ring and Shear Plate Connectors

12.3.1 Terminology

12.3.1.1 "Edge distance" is the distance from the edge of a member to the center of the nearest split ring or shear plate connector, measured perpendicular to grain. When a member is loaded perpendicular to grain, the loaded edge shall be defined as the edge toward which the load is acting. The unloaded edge shall be defined as the edge opposite the loaded edge (see Figure 12J).

12.3.1.2 "End distance" is the distance measured parallel to grain from the square-cut end of a member to the center of the nearest split ring or shear plate connector (see Figure 12J). If the end of a member is not cut at a right angle to its longitudinal axis, the end distance, measured parallel to the longitudinal axis from any point on the center half of the transverse connector diameter, shall not be less than the end distance required for a square-cut member. In no case shall the perpendicular distance from the center of a connector to the sloping end cut of a member, be less than the required edge distance (see Figure 12K).

Figure 12J Connection Geometry for Split Rings and Shear Plates

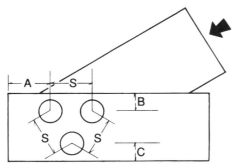

A = End Distance
B = Unloaded Edge Distance
C = Loaded Edge Distance
S = Spacing

12.3.1.3 "Spacing" is the distance between centers of split ring or shear plate connectors measured along a line joining their centers (see Figure 12J).

Figure 12K End Distance for Members with Sloping End Cut

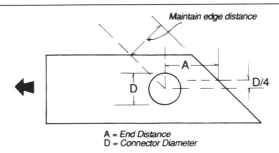

A = End Distance
D = Connector Diameter

12.3.2 Geometry Factor, C_Δ

Reference design values are for split ring and shear plate connectors with edge distance, end distance, and spacing greater than or equal to the minimum required for $C_\Delta = 1.0$. When the edge distance, end distance, or spacing provided is less than the minimum required $C_\Delta = 1.0$, reference design values shall be multiplied by the smallest applicable geometry factor, C_Δ, determined from the edge distance, end distance, and spacing requirements for split ring and shear plate connectors (see 12.3.3, 12.3.4, and 12.3.5). The smallest geometry factor for any split ring or shear plate connector in a group shall apply to all split ring and shear plate connectors in the group.

12.3.3 Edge Distance

12.3.3.1 Members Loaded Parallel or Perpendicular to Grain. Minimum edge distances and associated geometry factors, C_Δ, for split ring and shear plate connectors installed in side grain and loaded parallel or perpendicular to grain are provided in Table 12.3. When the actual loaded edge distance is greater than or equal to the minimum loaded edge distance, but less than the minimum loaded edge distance for $C_\Delta = 1.0$, the geometry factor, C_Δ, shall be determined by linear interpolation.

12.3.3.2 Members Loaded at Angle to Grain. When members are loaded at an angle to grain, θ, other than 0° or 90°, the minimum loaded edge distances and the minimum unloaded edge distances in Table 12.3 shall apply for all angles of load to grain. Minimum loaded

12

edge distances for $C_\Delta = 1.0$ shall be determined as follows:

(a) When $45° \leq \theta \leq 90°$, the minimum loaded edge distance for $C_\Delta = 1.0$ for perpendicular to grain loading shall apply.

(b) When $0° \leq \theta < 45°$, the minimum loaded edge distance for $C_\Delta = 1.0$ shall be determined by linear interpolation between the minimum loaded edge distance and the minimum loaded edge distance for $C_\Delta = 1.0$ for perpendicular to grain loading.

When a member is loaded at an angle to grain, θ, other than 0° or 90°, the geometry factor, C_Δ, based on edge distance requirements shall apply to both the reference parallel and perpendicular to grain design values (P, Q).

12.3.4 End Distance

12.3.4.1 Members Loaded Parallel or Perpendicular to Grain. Minimum end distances and associated geometry factors, C_Δ, for split ring and shear plate connectors installed in side grain and loaded parallel or perpendicular to grain are provided in Table 12.3. When the actual end distance is greater than or equal to the minimum end distance, but less than the minimum end distance for $C_\Delta = 1.0$, the geometry factor, C_Δ, shall be determined by linear interpolation.

12.3.4.2 Members Loaded at Angle to Grain. When members are loaded at an angle to grain, θ, other than 0° or 90°, minimum end distances and minimum end distances for $C_\Delta = 1.0$ shall be determined by linear interpolation between tabulated end distances for parallel and perpendicular to grain loading.

12.3.5 Spacing

12.3.5.1 Members Loaded Parallel or Perpendicular to Grain. Minimum parallel and perpendicular to grain spacings and associated geometry factors, C_Δ, for split ring and shear plate connectors installed in side grain and loaded parallel or perpendicular to grain are provided in Table 12.3. When the line joining the centers of two adjacent split ring or shear plate connectors is at an angle to grain other than 0° or 90°, the minimum spacing and the minimum spacing for $C_\Delta = 1.0$ shall be determined in accordance with the graphical method specified in References 47 and 48. When the actual spacing between split ring or shear plate connectors is greater than the minimum spacing, but less than the minimum spacing for $C_\Delta = 1.0$, the geometry factor, C_Δ, shall be determined by linear interpolation.

12.3.5.2 Members Loaded at Angle to Grain. When members are loaded at an angle to grain, θ, other than 0° or 90°, the minimum spacing and minimum spacing for $C_\Delta = 1.0$ shall be determined in accordance with the graphical method specified in References 50 and 52.

12.3.6 Split Ring and Shear Plate Connectors in End Grain

12.3.6.1 The provisions for edge distance, end distance, and spacing given in 12.3.3, 12.3.4, and 12.3.5 for split ring and shear plate connectors installed in side grain shall apply to split ring and shear plate connectors installed in square-cut surfaces and sloping surfaces as follows (see 12.2.6 for definitions and terminology):

(a) Square-cut surface, loaded in any direction - apply provisions for perpendicular to grain loading.

(b) Sloping surface with α from 45° to 90°, loaded in any direction - apply provisions for perpendicular to grain loading.

(c) Sloping surface with α less than 45°, loaded parallel to axis of cut - apply provisions for parallel to grain loading.

(d) Sloping surface with α less than 45°, loaded perpendicular to axis of cut - apply provisions for perpendicular to grain loading.

(e) Sloping surface with α less than 45°, loaded at angle ϕ to axis of cut - apply provisions for members loaded at angles to grain other than 0° or 90°.

12.3.6.2 When split ring or shear plate connectors are installed in end grain, the members shall be designed for shear parallel to grain in accordance with 3.4.3.3.

12.3.7 Multiple Split Ring or Shear Plate Connectors

12.3.7.1 When a connection contains two or more split ring or shear plate connector units which are in the same shear plane, are aligned in the direction of load, and on separate bolts or lag screws, the group action factor, C_g, shall be as specified in 10.3.6 and the total adjusted design value for the connection shall be as specified in 10.2.2.

12.3.7.2 If grooves for two sizes of split rings are cut concentric in the same wood surface, split ring connectors shall be installed in both grooves and the reference design value shall be taken as the reference design value for the larger split ring connector.

12.3.7.3 Local stresses in connections using multiple fasteners shall be evaluated in accordance with principles of engineering mechanics (see 10.1.2).

SPLIT RING AND SHEAR PLATE CONNECTORS

12

Table 12.3 Geometry Factors, C_Δ, for Split Ring and Shear Plate Connectors

		2-1/2" Split Ring Connectors & 2-5/8" Shear Plate Connectors				4" Split Ring Connectors & 4" Shear Plate Connectors			
		Parallel to grain loading		Perpendicular to grain loading		Parallel to grain loading		Perpendicular to grain loading	
		Minimum Value	Minimum for $C_\Delta = 1.0$	Minimum Value	Minimum for $C_\Delta = 1.0$	Minimum Value	Minimum for $C_\Delta = 1.0$	Minimum Value	Minimum for $C_\Delta = 1.0$
Edge Distance	Unloaded Edge C_Δ	1-3/4" / 1.0	1-3/4" / 1.0	1-3/4" / 1.0	1-3/4" / 1.0	2-3/4" / 1.0	2-3/4" / 1.0	2-3/4" / 1.0	2-3/4" / 1.0
	Loaded Edge C_Δ	1-3/4" / 1.0	1-3/4" / 1.0	1-3/4" / 0.83	2-3/4" / 1.0	2-3/4" / 1.0	2-3/4" / 1.0	2-3/4" / 0.83	3-3/4" / 1.0
End Distance	Tension Member C_Δ	2-3/4" / 0.625	5-1/2" / 1.0	2-3/4" / 0.625	5-1/2" / 1.0	3-1/2" / 0.625	7" / 1.0	3-1/2" / 0.625	7" / 1.0
	Compression Member C_Δ	2-1/2" / 0.625	4" / 1.0	2-3/4" / 0.625	5-1/2" / 1.0	3-1/4" / 0.625	5-1/2" / 1.0	3-1/2" / 0.625	7" / 1.0
Spacing	Spacing parallel to grain C_Δ	3-1/2" / 0.5	6-3/4" / 1.0	3-1/2" / 1.0	3-1/2" / 1.0	5" / 0.5	9" / 1.0	5" / 1.0	5" / 1.0
	Spacing perpendicular to grain C_Δ	3-1/2" / 1.0	3-1/2" / 1.0	3-1/2" / 0.5	4-1/4" / 1.0	5" / 1.0	5" / 1.0	5" / 0.5	6" / 1.0

TIMBER RIVETS

13

13.1 General

Design criteria for timber rivet joints apply to timber rivets that satisfy the requirements of 13.1.1 loaded in single shear, with steel side plates on Douglas Fir-Larch or Southern Pine structural glued laminated timber manufactured in accordance with ANSI/AITC A190.1.

13.1.1 Quality of Rivets and Steel Side Plates

13.1.1.1 Design provisions and reference design values herein apply to timber rivets that are hot-dip galvanized in accordance with ASTM A 153 and manufactured from AISI 1035 steel to have the following properties tested in accordance with ASTM A 370:

Hardness	Ultimate tensile strength, F_u
Rockwell C32-39	145,000 psi, minimum

See Appendix M for rivet dimensions.

13.1.1.2 Steel side plates shall conform to ASTM Standard A 36 with a minimum 1/8" thickness. See Appendix M for steel side plate dimensions.

13.1.1.3 For wet service conditions, steel side plates shall be hot-dip galvanized in accordance with ASTM A 153.

13.1.2 Fabrication and Assembly

13.1.2.1 Each rivet shall, in all cases, be placed with its major cross-sectional dimension aligned paral-

lel to the grain. Design criteria are based on rivets driven through circular holes in the side plates until the conical heads are firmly seated, but rivets shall not be driven flush. (Timber rivets at the perimeter of the group shall be driven first. Successive timber rivets shall be driven in a spiral pattern from the outside to the center of the group.)

13.1.2.2 The maximum penetration of any rivet shall be 70% of the thickness of the wood member. Except as permitted by 13.1.2.3, for joints with rivets driven from opposite faces of a wood member, the rivet length shall be such that the points do not overlap.

13.1.2.3 For joints where rivets are driven from opposite faces of a wood member such that their points overlap, the minimum spacing requirements of 13.3.1 shall apply to the distance between the rivets at their points and the maximum penetration requirement of 13.1.2.2 shall apply. The reference lateral design value of the connection shall be calculated in accordance with 13.2 considering the connection to be a one sided timber rivet joint, with:

(a) the number of rivets associated with the one plate equalling the total number of rivets at the joint, and

(b) s_p and s_q determined as the distances between the rivets at their points.

13.2 Reference Design Values

13.2.1 Parallel to Grain Loading

For timber rivet connections (one plate and rivets associated with it) where:

(a) the load acts perpendicular to the axis of the timber rivets

(b) member thicknesses, edge distances, end distances, and spacing are sufficient to develop full adjusted design values (see 13.3)

(c) timber rivets are installed in the side grain of wood members the reference design value per rivet joint parallel to grain, P, shall be calculated as the lesser of reference rivet capacity, P_r, and reference wood capacity, P_w:

$$P_r = 280 \, p^{0.32} \, n_R \, n_C \qquad (13.2\text{-}1)$$

P_w = reference wood capacity design values parallel to grain (Tables 13.2.1A through 13.2.1F) using wood member thickness for the member dimension in Tables 13.2.1A through 13.2.1F for connections with steel plates on opposite sides; and twice the wood member thickness for the member dimension in Tables 13.2.1A through 13.2.1F for connections having only one plate, lbs.

where:

p = depth of penetration of rivet in wood member (see Appendix M), in.

= rivet length – plate thickness – 1/8"

n_R = number of rows of rivets parallel to direction of load

n_c = number of rivets per row

Reference design values, P, for timber rivet connections parallel to grain shall be multiplied by all applicable adjustment factors (see Table 10.3.1) to obtain adjusted design values, P'.

13.2.2 Perpendicular to Grain Loading

For timber rivet connections (one plate and rivets associated with it) where:
(a) the load acts perpendicular to the axis of the timber rivets
(b) member thicknesses, edge distances, end distances, and spacing are sufficient to develop full adjusted design values (see 13.3)
(c) timber rivets are installed in the side grain of wood members the reference design value per rivet joint perpendicular to grain, Q, shall be calculated as the lesser of reference rivet capacity, Q_r, and reference wood capacity, Q_w.

$$Q_r = 160 \, p^{0.32} n_R n_c \qquad (13.2\text{-}2)$$

$$Q_w = q_w p^{0.8} C_\Delta \qquad (13.2\text{-}3)$$

where:

p = depth of penetration of rivet in wood member (see Appendix M), in.

= rivet length – plate thickness – 1/8"

n_R = number of rows of rivets parallel to direction of load

n_c = number of rivets per row

q_w = value determined from Table 13.2.2A, lbs.

C_Δ = geometry factor determined from Table 13.2.2B

Reference design values, Q, for timber rivet connections perpendicular to grain shall be multiplied by all applicable adjustment factors (see Table 10.3.1) to obtain adjusted design values, Q'.

13.2.3 Metal Side Plate Factor, C$_{st}$

The reference design value parallel to grain, P, or perpendicular to grain, Q, for timber rivet connections, when reference rivet capacity (P_r, Q_r) controls, shall be multiplied by the appropriate metal side plate factor, C_{st}, specified in Table 13.2.3:

Table 13.2.3 Metal Side Plate Factor, C$_{st}$, for Timber Rivet Connections

Metal Side Plate Thickness, t$_s$	C$_{st}$
$t_s \geq 1/4$"	1.00
$3/16" \leq t_s < 1/4"$	0.90
$1/8" \leq t_s < 3/16"$	0.80

13.2.4 Load at Angle to Grain

When a load acts in the plane of the wood surface at an angle, θ, to grain other than 0° or 90°, the adjusted design value, N', for a timber rivet connection shall be determined as follows (see Appendix J):

$$N' = \frac{P'Q'}{P' \sin^2 \theta + Q' \cos^2 \theta} \qquad (13.2\text{-}4)$$

13.2.5 Timber Rivets in End Grain

When timber rivets are used in end grain, the factored lateral resistance of the joint shall be 50% of that for perpendicular to side grain applications when the slope of cut is 90° to the side grain. For sloping end cuts, these values can be increased linearly to 100% of the applicable parallel or perpendicular to side grain value.

13.2.6 Design of Metal Parts

Metal parts shall be designed in accordance with applicable metal design procedures (see 10.2.3).

TIMBER RIVETS

13

13.3 Placement of Timber Rivets

13.3.1 Spacing Between Rivets

Minimum spacing of rivets shall be 1/2" perpendicular to grain, s_q, and 1" parallel to grain, s_p.

13.3.2 End and Edge Distance

Minimum values for end distance (a_p, a_q) and edge distance (e_p, e_q) as shown and noted in Figure 13A, are listed in Table 13.3.2.

Table 13.3.2 Minimum End and Edge Distances for Timber Rivet Joints

Number of rivet rows, n_R	Minimum end distance, a, in.		Minimum edge distance, e, in.	
	Load Parallel to grain, a_p	Load perpendicular to grain, a_q	Unloaded Edge e_p	Loaded edge e_q
1, 2	3	2	1	2
3 to 8	3	3	1	2
9, 10	4	3-1/8	1	2
11, 12	5	4	1	2
13, 14	6	4-3/4	1	2
15, 16	7	5-1/2	1	2
17 and greater	8	6-1/4	1	2

Note: End and edge distance requirements are shown in Figure 13A.

Figure 13A End and Edge Distance Requirements for Timber Rivet Joints

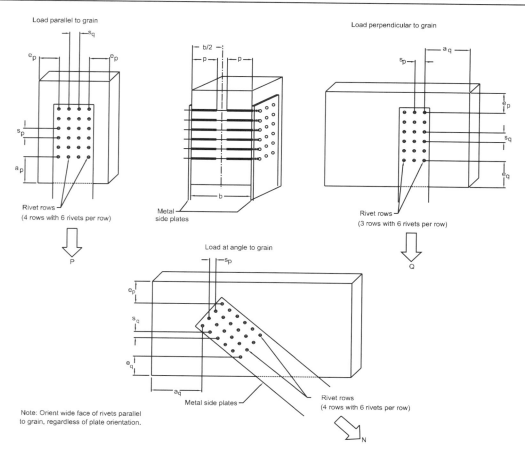

Table 13.2.1A Reference Wood Capacity Design Values Parallel to Grain, P_w, for Timber Rivets

Rivet Length = 1-1/2" $s_p = 1$" $s_q = 1$"

Member Thickness in.	Rivets per row	P_w (lbs.) No. of rows per side									
		2	4	6	8	10	12	14	16	18	20
3	2	2050	4900	7650	10770	14100	17050	19760	22660	25690	28990
	4	3010	6460	9700	13530	17450	20840	23870	27020	30530	34460
	6	4040	8010	11770	16320	20870	24770	27950	31450	35710	40300
	8	5110	9480	13970	18840	23910	28230	31990	35760	40130	45290
	10	5900	10930	15880	21390	26940	32020	35660	40080	44830	50590
	12	6670	12100	17760	23980	29980	35010	39780	44480	49570	55940
	14	7310	13540	19400	26380	32740	38610	43090	48640	54720	61750
	16	7670	14960	21380	28260	35470	41670	46310	52870	59350	66970
	18	8520	16250	23290	30440	38010	44500	50050	56120	63840	70970
	20	9030	17770	24950	32300	40160	46880	52590	59800	66880	74300
5	2	2680	5160	5980	7250	9280	10860	12470	15150	19410	24260
	4	3930	6610	7610	9050	11460	13390	15110	17890	22090	26280
	6	5280	8190	9290	10890	13770	15870	18080	21120	25640	29870
	8	6690	9700	10940	12740	15950	18230	20580	23780	28450	32570
	10	7720	11160	12550	14550	18120	20600	23140	26550	31500	36850
	12	8730	12680	14170	16240	20100	23100	25410	29610	35000	40730
	14	9560	14160	15720	17980	22210	25460	27940	32450	38220	44250
	16	10030	15610	17330	19650	24200	27680	30320	35100	42230	48910
	18	11150	17020	18770	21450	26110	29780	33140	38370	46160	51900
	20	11800	18410	20310	23000	28270	32260	35900	41570	50030	56260
6.75	2	2930	4810	5550	6740	8630	10110	11610	14120	18080	22630
	4	4300	6170	7080	8420	10680	12490	14100	16700	20630	24570
	6	5780	7650	8640	10150	12840	14820	16890	19740	23980	27960
	8	7320	9060	10190	11880	14890	17040	19250	22240	26630	30510
	10	8440	10420	11690	13580	16920	19260	21640	24850	29500	34540
	12	9540	11850	13210	15150	18780	21610	23780	27730	32800	38200
	14	10450	13230	14650	16790	20760	23820	26170	30410	35830	41520
	16	10970	14590	16160	18350	22630	25910	28410	32900	39610	45910
	18	12190	15910	17510	20040	24420	27890	31050	35980	43310	48740
	20	12910	17210	18950	21490	26450	30210	33650	38990	46950	52860
8.5 and greater	2	2930	4740	5460	6630	8500	9950	11440	13900	17810	22290
	4	4300	6080	6970	8290	10520	12300	13890	16450	20330	24210
	6	5780	7530	8510	10000	12650	14600	16640	19460	23630	27560
	8	7320	8920	10030	11700	14670	16790	18970	21930	26250	30080
	10	8440	10270	11520	13370	16680	18980	21330	24500	29090	34060
	12	9540	11670	13010	14930	18510	21300	23450	27340	32340	37670
	14	10450	13040	14430	16540	20460	23480	25800	29980	35340	40960
	16	10970	14370	15920	18080	22310	25540	28010	32450	39060	45290
	18	12190	15670	17250	19750	24070	27490	30620	35480	42720	48090
	20	12910	16950	18670	21180	26070	29790	33190	38460	46310	52150

Note: Member dimension is identified as "b" in Figure 13A for connections with steel side plates on opposite sides. For connections having only one plate, member dimension is twice the thickness of the wood member. Linear interpolation for intermediate values shall be permitted.

TIMBER RIVETS

13

Table 13.2.1B   Reference Wood Capacity Design Values Parallel to Grain, P_w, for Timber Rivets

Rivet Length = 1-1/2" s_p = 1-1/2" s_q = 1"

Member Thickness in.	Rivets per row	P_w (lbs.) No. of rows per side									
		2	4	6	8	10	12	14	16	18	20
3	2	2320	5650	8790	12270	16000	19800	23200	26100	29360	33180
	4	3420	7450	11150	15420	19810	24200	28020	31130	34900	39430
	6	4580	9230	13530	18600	23690	28760	32810	36230	40810	46120
	8	5810	10920	16060	21480	27150	32780	37550	41200	45860	51830
	10	6700	12600	18250	24380	30590	37180	41870	46180	51230	57890
	12	7570	13940	20420	27340	34040	40650	46700	51250	56650	64020
	14	8290	15600	22310	30070	37180	44840	50590	56040	62540	70670
	16	8710	17250	24580	32220	40280	48400	54360	60910	67820	76650
	18	9680	18720	26770	34700	43150	51680	58750	64660	72960	81220
	20	10250	20480	28680	36820	45600	54450	61740	68900	76440	85030
5	2	3040	5360	6740	8600	11930	14870	18310	23450	32100	42850
	4	4470	7660	9560	11970	16430	20450	24740	30870	40740	51580
	6	5990	9910	12180	15050	20610	25320	30910	38070	49400	60320
	8	7590	12000	14680	18020	24440	29760	36020	43870	56110	67790
	10	8760	14010	17090	20880	28170	34120	41080	49700	63030	75720
	12	9900	16080	19480	23530	31570	38650	45570	55990	70740	83740
	14	10850	18080	21770	26240	35120	42890	50480	61810	77820	92440
	16	11390	20040	24140	28830	38490	46900	55080	67230	86450	100250
	18	12660	21950	26250	31620	41690	50680	60450	73800	94910	106230
	20	13400	23810	28500	34010	45310	55090	65720	80250	99970	111210
6.75	2	3320	5000	6260	8000	11110	13850	17060	21870	29940	39990
	4	4890	7150	8900	11150	15330	19090	23110	28850	38090	48440
	6	6560	9250	11340	14040	19240	23660	28900	35620	46240	57570
	8	8310	11210	13680	16810	22840	27840	33710	41080	52570	64320
	10	9580	13090	15930	19500	26330	31930	38470	46570	59100	73900
	12	10830	15020	18170	21980	29520	36180	42700	52490	66360	82550
	14	11860	16900	20310	24520	32860	40180	47320	57980	73030	90400
	16	12460	18730	22520	26950	36030	43940	51650	63090	81170	100510
	18	13840	20520	24500	29560	39040	47500	56710	69290	89150	107180
	20	14660	22270	26610	31810	42440	51650	61680	75360	96980	116640
8.5 and greater	2	3320	4930	6160	7880	10930	13640	16810	21540	29490	39400
	4	4890	7050	8760	10990	15100	18800	22770	28430	37540	47750
	6	6560	9110	11170	13830	18960	23310	28490	35110	45590	56770
	8	8310	11040	13480	16560	22510	27440	33230	40510	51840	63430
	10	9580	12890	15690	19210	25960	31480	37930	45920	58280	72900
	12	10830	14800	17900	21660	29100	35670	42110	51770	65450	81440
	14	11860	16650	20000	24170	32390	39610	46670	57190	72040	89190
	16	12460	18450	22190	26560	35520	43330	50940	62240	80080	99190
	18	13840	20220	24140	29140	38490	46850	55940	68350	87960	105780
	20	14660	21940	26220	31360	41840	50940	60840	74350	95690	115120

Note: Member dimension is identified as "b" in Figure 13A for connections with steel side plates on opposite sides. For connections having only one plate, member dimension is twice the thickness of the wood member. Linear interpolation for intermediate values shall be permitted.

Table 13.2.1C Reference Wood Capacity Design Values Parallel to Grain, P_w, for Timber Rivets

Rivet Length = 2-1/2" s_p = 1" s_q = 1"

Member Thickness in.	Rivets per row	P_w (lbs.) No. of rows per side									
		2	4	6	8	10	12	14	16	18	20
5	2	2340	5610	8750	12310	16120	19500	22600	25910	29380	33160
	4	3440	7390	11100	15470	19950	23830	27290	30900	34920	39400
	6	4620	9160	13460	18660	23860	28320	31970	35960	40830	46080
	8	5850	10840	15980	21550	27350	32280	36580	40900	45890	51790
	10	6750	12500	18160	24460	30810	36610	40780	45840	51260	57850
	12	7630	13830	20310	27420	34280	40030	45490	50870	56690	63970
	14	8360	15480	22190	30170	37450	44150	49280	55620	62580	70620
	16	8770	17110	24450	32320	40570	47660	52960	60450	67870	76590
	18	9750	18580	26630	34810	43460	50890	57230	64170	73010	81160
	20	10320	20320	28530	36940	45920	53610	60140	68380	76480	84960
6.75	2	2710	6490	10130	14260	18660	22570	26170	30000	34020	38390
	4	3980	8550	12850	17910	22580	26120	29190	34220	40420	45620
	6	5350	10600	15590	20390	25510	29030	32670	37760	45400	52330
	8	6770	12550	18500	22880	28260	31840	35470	40500	47980	54310
	10	7810	14480	21020	25280	30980	34680	38400	43540	51130	59140
	12	8830	16020	23510	27430	33360	37720	40900	47070	55050	63330
	14	9670	17920	25690	29640	35930	40500	43810	50240	58540	67000
	16	10160	19810	28310	31700	38300	43040	46460	53110	63200	72360
	18	11290	21510	30160	33950	40490	45390	49750	56870	67670	75240
	20	11950	23530	32140	35770	43070	48280	52920	60500	72000	80080
8.5	2	3070	7350	10580	13060	16620	19300	21990	26530	33760	41900
	4	4510	9690	12400	14710	18410	21240	23720	27810	34060	40180
	6	6060	12000	14390	16700	20790	23640	26610	30780	37040	42750
	8	7670	13920	16320	18720	23050	25970	28960	33100	39250	44510
	10	8850	15730	18150	20680	25290	28330	31420	35660	41930	48600
	12	10010	17590	19970	22430	27270	30870	33520	38630	45240	52180
	14	10960	19360	21660	24250	29400	33190	35960	41310	48200	55320
	16	11510	21050	23410	25950	31370	35320	38200	43740	52130	59860
	18	12790	22670	24900	27810	33200	37290	40960	46920	55920	62350
	20	13540	24220	26510	29310	35350	39720	43640	49990	59580	66480
10.5	2	3400	7730	9830	11980	15210	17650	20110	24260	30870	38340
	4	5000	9490	11460	13490	16860	19460	21740	25500	31230	36880
	6	6710	11400	13250	15310	19060	21690	24430	28270	34030	39320
	8	8490	13150	15020	17170	21150	23850	26610	30440	36110	41000
	10	9800	14810	16700	18980	23230	26040	28900	32840	38630	44830
	12	11080	16520	18360	20600	25060	28400	30870	35610	41730	48190
	14	12130	18140	19910	22280	27040	30560	33150	38110	44500	51140
	16	12740	19680	21520	23850	28870	32550	35240	40390	48170	55390
	18	14160	21160	22900	25570	30570	34390	37820	43350	51710	57750
	20	14990	22580	24380	26970	32570	36640	40310	46220	55140	61620
12.5 and greater	2	3540	7610	9540	11590	14710	17060	19440	23450	29840	37060
	4	5210	9300	11100	13040	16300	18820	21030	24670	30230	35700
	6	6990	11140	12840	14810	18440	20990	23650	27370	32960	38100
	8	8860	12840	14540	16620	20470	23090	25780	29490	35000	39750
	10	10220	14440	16160	18370	22490	25230	28010	31830	37450	43490
	12	11550	16090	17770	19940	24270	27520	29920	34530	40470	46760
	14	12650	17650	19270	21580	26190	29620	32150	36970	43180	49650
	16	13290	19140	20840	23100	27970	31560	34180	39190	46760	53800
	18	14760	20570	22170	24770	29630	33350	36690	42080	50210	56110
	20	15630	21940	23600	26130	31570	35550	39120	44880	53560	59880

TIMBER RIVETS

13

Note: Member dimension is identified as "b" in Figure 13A for connections with steel side plates on opposite sides. For connections having only one plate, member dimension is twice the thickness of the wood member. Linear interpolation for intermediate values shall be permitted.

Table 13.2.1D   Reference Wood Capacity Design Values Parallel to Grain, P_w, for Timber Rivets

Rivet Length = 2-1/2" s_p = 1-1/2" s_q = 1"

Member Thickness in.	Rivets per row	P_w (lbs.) No. of rows per side									
		2	4	6	8	10	12	14	16	18	20
5	2	2660	6460	10050	14040	18300	22640	26530	29850	33580	37950
	4	3910	8520	12750	17640	22650	27670	32040	35600	39900	45090
	6	5240	10560	15480	21270	27090	32890	37530	41430	46670	52740
	8	6640	12490	18370	24560	31050	37490	42940	47120	52450	59270
	10	7660	14410	20870	27880	34980	42520	47880	52810	58580	66200
	12	8660	15950	23350	31260	38920	46490	53400	58610	64790	73210
	14	9480	17840	25510	34390	42520	51270	57850	64080	71520	80820
	16	9960	19720	28110	36850	46060	55340	62170	69650	77560	87650
	18	11070	21410	30610	39680	49350	59090	67180	73940	83440	92880
	20	11720	23420	32800	42110	52140	62260	70600	78790	87410	97230
6.75	2	3070	7480	11640	16250	21190	26210	30720	34560	38880	43930
	4	4520	9860	14770	20420	26230	32040	37100	41220	46200	52210
	6	6070	12220	17920	24630	31370	38080	43450	47970	54030	61060
	8	7690	14460	21260	28440	35950	43400	49720	54550	60720	68620
	10	8870	16690	24160	32280	40500	49230	55430	61150	67820	76650
	12	10030	18460	27030	36200	45060	53820	61830	67860	75010	84760
	14	10980	20660	29530	39820	49220	59360	66980	74190	82800	93570
	16	11530	22830	32550	42660	53320	64070	71980	80640	89800	101480
	18	12810	24790	35440	45940	57130	68420	77780	85600	96600	107530
	20	13560	27110	37970	48750	60370	72080	81740	91220	101200	112570
8.5	2	3480	8230	11610	14990	20600	25440	31030	39170	44060	49790
	4	5120	11170	14980	18590	25140	30870	36920	45610	52360	59170
	6	6880	13850	18020	21920	29500	35710	43060	52490	61230	69190
	8	8710	16390	20820	25060	33380	40030	47840	57640	68810	77760
	10	10050	18910	23430	28020	37080	44230	52570	62910	76860	86860
	12	11360	20920	25960	30640	40320	48610	56590	68770	85000	96060
	14	12440	23410	28300	33320	43740	52600	61110	74030	92320	106040
	16	13070	25860	30710	35810	46880	56250	65240	78780	100360	115000
	18	14520	27900	32770	38510	49810	59620	70230	84840	108100	121860
	20	15370	29860	34970	40700	53190	63690	75060	90690	114690	127580
10.5	2	3860	7930	10760	13740	18860	23280	28400	36090	48770	55110
	4	5670	10740	13810	17050	23050	28310	33880	41870	54800	65490
	6	7610	13360	16580	20110	27080	32800	39590	48290	62110	76590
	8	9640	15700	19140	23010	30670	36830	44050	53110	67340	81680
	10	11130	17870	21540	25740	34110	40740	48470	58060	72990	90530
	12	12580	20050	23860	28170	37130	44820	52230	63540	79580	98220
	14	13770	22110	26020	30660	40300	48540	56460	68470	85450	104980
	16	14460	24060	28240	32970	43240	51950	60330	72930	92990	114320
	18	16070	25920	30140	35470	45960	55110	65010	78610	100260	119710
	20	17020	27710	32170	37520	49120	58920	69530	84110	107290	128200
12.5 and greater	2	4020	7800	10440	13290	18230	22500	27450	34890	47430	57470
	4	5920	10500	13370	16490	22300	27390	32790	40540	53060	66920
	6	7940	13040	16050	19460	26210	31770	38350	46780	60200	74320
	8	10060	15300	18530	22270	29710	35680	42690	51500	65300	79260
	10	11600	17390	20850	24930	33050	39490	47000	56320	70830	87900
	12	13120	19490	23110	27290	35980	43460	50680	61670	77270	95420
	14	14370	21480	25200	29700	39080	47090	54800	66480	83000	102030
	16	15080	23370	27350	31950	41930	50420	58580	70850	90360	111160
	18	16760	25170	29200	34380	44590	53500	63140	76390	97460	116450
	20	17750	26900	31170	36380	47660	57220	67550	81760	104330	124740

Note: Member dimension is identified as "b" in Figure 13A for connections with steel side plates on opposite sides. For connections having only one plate, member dimension is twice the thickness of the wood member. Linear interpolation for intermediate values shall be permitted.

Table 13.2.1E   Reference Wood Capacity Design Values Parallel to Grain, P_w, for Timber Rivets

Rivet Length = 3-1/2" $s_p = 1"$ $s_q = 1"$

Member Thickness in.	Rivets per row	P_w (lbs.) No. of rows per side									
		2	4	6	8	10	12	14	16	18	20
6.75	2	2440	5850	9130	12850	16820	20350	23590	27040	30670	34610
	4	3590	7710	11580	16150	20820	24870	28490	32250	36450	41130
	6	4820	9560	14050	19480	24910	29560	33370	37540	42620	48100
	8	6100	11310	16680	22490	28550	33700	38180	42690	47900	54060
	10	7040	13050	18950	25530	32160	38220	42570	47850	53510	60380
	12	7960	14440	21200	28630	35780	41780	47480	53100	59170	66770
	14	8720	16160	23160	31490	39090	46090	51440	58050	65320	73710
	16	9160	17860	25530	33740	42340	49740	55280	63100	70840	79940
	18	10170	19390	27790	36330	45370	53120	59740	66990	76210	84710
	20	10770	21210	29780	38560	47930	55960	62770	71380	79830	88680
8.5	2	2710	6490	10130	14250	18660	22570	26160	29990	34010	38380
	4	3980	8550	12840	17910	23090	27580	31600	35770	40420	45610
	6	5350	10600	15590	21600	27620	32790	37000	41630	47270	53350
	8	6770	12550	18500	24940	31660	37370	42350	47340	53120	59950
	10	7810	14480	21020	28320	35670	42390	47210	53060	59340	66970
	12	8830	16020	23510	31750	39680	46340	52660	58890	65620	74060
	14	9670	17920	25690	34920	43350	51110	57050	64390	72440	81750
	16	10160	19810	28310	37420	46960	55170	61310	69980	78560	88660
	18	11280	21510	30830	40300	50310	58910	66250	74290	84510	93950
	20	11950	23520	33030	42760	53160	62060	69620	79160	88540	98360
10.5	2	3020	7240	11300	15900	20820	25180	29190	33460	37940	42820
	4	4440	9540	14330	19980	25760	30770	35250	39900	45090	50890
	6	5960	11830	17390	24100	30820	36580	41280	46440	52740	59510
	8	7550	14000	20630	27830	35320	40570	44460	49980	58370	65230
	10	8720	16150	23450	31420	38000	41760	45450	50740	58770	67160
	12	9850	17870	26230	32850	39220	43470	46330	52530	60650	68990
	14	10790	19990	28660	34370	40770	45050	47920	54190	62370	70660
	16	11330	22100	31580	35730	42190	46510	49400	55710	65540	74330
	18	12590	23990	33750	37340	43510	47850	51650	58300	68630	75640
	20	13330	26240	35340	38490	45290	49850	53850	60830	71650	79050
12.5	2	3320	7960	12420	17490	22890	27690	32090	36790	41720	47090
	4	4890	10490	15760	21970	28330	33840	38760	43880	49590	55960
	6	6560	13010	19120	25230	31370	35100	38780	44070	52180	59340
	8	8310	15390	22350	26580	32170	35480	38780	43560	50870	56880
	10	9580	17760	24250	27850	33280	36450	39640	44260	51300	58690
	12	10830	19650	25950	28920	34280	37940	40440	45890	53030	60430
	14	11870	21990	27400	30150	35610	39340	41890	47420	54640	62020
	16	12460	24300	28890	31290	36860	40660	43240	48840	57520	65380
	18	13840	26360	30040	32670	38030	41880	45280	51190	60340	66660
	20	14660	28020	31320	33670	39620	43680	47270	53490	63100	69790
14.5 and greater	2	3580	8580	13390	18850	24670	29840	34590	39650	44970	50750
	4	5270	11020	16940	22830	29290	33640	37040	42730	51500	59860
	6	7070	13590	19540	23990	29520	32900	36290	41210	48800	55490
	8	8950	15930	21540	25060	30160	33200	36280	40760	47610	53260
	10	10330	18090	23150	26150	31160	34110	37110	41450	48060	55040
	12	11680	20230	24620	27120	32090	35530	37890	43020	49740	56740
	14	12790	22170	25890	28250	33350	36870	39280	44500	51310	58300
	16	13430	23950	27220	29310	34540	38120	40580	45870	54070	61530
	18	14920	25580	28250	30610	35650	39300	42530	48120	56770	62800
	20	15800	27070	29430	31550	37160	41020	44440	50330	59420	65810

Note: Member dimension is identified as "b" in Figure 13A for connections with steel side plates on opposite sides. For connections having only one plate, member dimension is twice the thickness of the wood member. Linear interpolation for intermediate values shall be permitted.

TIMBER RIVETS

13

Table 13.2.1F Reference Wood Capacity Design Values Parallel to Grain, P_w, for Timber Rivets

Rivet Length = 3-1/2" s_p = 1-1/2" s_q = 1"

Member Thickness in.	Rivets per row	P_w (lbs.) No. of rows per side									
		2	4	6	8	10	12	14	16	18	20
6.75	2	2770	6740	10490	14650	19100	23630	27690	31160	35050	39610
	4	4080	8890	13310	18410	23640	28880	33440	37160	41650	47070
	6	5470	11020	16160	22200	28280	34330	39170	43250	48710	55050
	8	6930	13040	19170	25640	32410	39130	44820	49180	54740	61860
	10	8000	15040	21780	29110	36510	44380	49970	55130	61150	69100
	12	9040	16640	24370	32630	40630	48520	55740	61180	67620	76420
	14	9900	18630	26630	35900	44380	53520	60390	66890	74650	84360
	16	10390	20590	29340	38460	48080	57770	64890	72710	80960	91490
	18	11550	22350	31950	41420	51510	61680	70130	77180	87090	96950
	20	12230	24450	34230	43960	54430	64990	73690	82240	91240	101490
8.5	2	3070	7480	11640	16250	21190	26210	30710	34560	38870	43930
	4	4520	9860	14760	20420	26220	32030	37090	41210	46190	52200
	6	6070	12220	17920	24630	31360	38080	43440	47960	54020	61050
	8	7690	14460	21260	28440	35950	43400	49710	54550	60710	68610
	10	8870	16680	24160	32280	40500	49220	55420	61140	67820	76640
	12	10020	18460	27030	36190	45060	53820	61820	67850	75000	84750
	14	10980	20660	29530	39810	49220	59360	66970	74180	82790	93560
	16	11530	22830	32540	42660	53320	64070	71970	80630	89790	101460
	18	12810	24790	35440	45940	57130	68410	77770	85600	96590	107520
	20	13560	27110	37970	48750	60360	72070	81730	91210	101190	112560
10.5	2	3430	8340	12980	18130	23640	29240	34260	38550	43360	49000
	4	5040	11000	16470	22780	29250	35740	41380	45980	51530	58240
	6	6770	13630	19990	27470	34990	42480	48460	53510	60270	68110
	8	8570	16130	23720	31720	40100	48420	55460	60850	67730	76540
	10	9890	18610	26950	36010	45180	54910	61830	68210	75660	85500
	12	11180	20590	30150	40380	50270	60040	68970	75690	83670	94550
	14	12250	23040	32940	43530	54910	65690	74710	82760	92360	104370
	16	12860	25470	36300	45490	58120	68370	78030	89950	100170	113190
	18	14290	27650	39530	47750	60310	70840	82200	95490	107750	119950
	20	15130	30250	42360	49450	63130	74240	86230	101750	112880	125570
12.5	2	3770	8940	14280	19930	25990	32150	37680	42390	47680	53890
	4	5550	12090	18110	25050	32170	39300	45500	50560	56670	64040
	6	7440	14990	21980	30210	38480	46710	53290	58840	66270	74890
	8	9430	17740	26080	32640	42400	49720	58300	66910	74480	84170
	10	10880	20470	29450	34550	44560	52030	60770	71680	83190	94020
	12	12300	22640	31480	36220	46470	54910	62900	75440	92000	103970
	14	13470	25340	33270	38080	48780	57570	65900	78860	97350	114770
	16	14140	28010	35150	39800	50920	60030	68660	81990	103470	124470
	18	15710	30410	36640	41810	52910	62310	72460	86620	109420	129590
	20	16640	33260	38300	43320	55450	65400	76150	91130	115220	136640
14.5 and greater	2	4060	8940	15370	21480	28010	34650	40610	45690	51390	58080
	4	5980	12730	19520	26990	34670	42350	49040	54490	61080	69020
	6	8020	16160	23590	28890	37960	44900	53060	63410	71430	80720
	8	10160	19120	25880	30610	39690	46550	54610	64800	80270	90710
	10	11720	21820	27800	32370	41740	48760	56990	67280	83550	101330
	12	13250	24280	29620	33930	43560	51520	59070	70900	87820	107340
	14	14520	26450	31250	35680	45760	54070	61960	74220	91690	111670
	16	15240	28390	32980	37310	47800	56430	64630	77250	97570	119050
	18	16940	30160	34370	39220	49710	58630	68270	81700	103290	122520
	20	17930	31770	35920	40670	52140	61590	71820	86040	108880	129320

Note: Member dimension is identified as "b" in Figure 13A for connections with steel side plates on opposite sides. For connections having only one plate, member dimension is twice the thickness of the wood member. Linear interpolation for intermediate values shall be permitted.

Table 13.2.2A Values of q_w (lbs.) Perpendicular to Grain for Timber Rivets

$$s_p = 1''$$

s_q in.	Rivets per row	Number of rows				
		2	4	6	8	10
1	2	776	809	927	1089	1255
	3	768	806	910	1056	1202
	4	821	870	963	1098	1232
	5	874	923	1013	1147	1284
	6	959	1007	1094	1228	1371
	7	1048	1082	1163	1297	1436
	8	1173	1184	1256	1391	1525
	9	1237	1277	1345	1467	1624
	10	1318	1397	1460	1563	1752
	11	1420	1486	1536	1663	1850
	12	1548	1597	1628	1786	1970
	13	1711	1690	1741	1882	2062
	14	1924	1802	1878	1997	2170
	15	2042	1937	1963	2099	2298
	16	2182	2102	2063	2218	2449
	17	2350	2223	2178	2313	2541
	18	2553	2365	2313	2422	2644
	19	2524	2432	2407	2548	2762
	20	2497	2506	2514	2692	2897
1-1/2	2	1136	1097	1221	1414	1630
	3	1124	1093	1199	1371	1561
	4	1202	1180	1268	1426	1601
	5	1280	1251	1334	1490	1668
	6	1404	1366	1442	1595	1780
	7	1534	1467	1532	1685	1865
	8	1717	1606	1654	1806	1980
	9	1811	1731	1772	1905	2110
	10	1929	1894	1923	2030	2275
	11	2078	2016	2023	2159	2403
	12	2265	2166	2145	2319	2559
	13	2504	2292	2293	2444	2678
	14	2817	2444	2473	2593	2818
	15	2989	2627	2586	2725	2984
	16	3193	2850	2717	2880	3181
	17	3439	3014	2869	3004	3300
	18	3737	3207	3047	3146	3434
	19	3695	3298	3171	3309	3588
	20	3655	3398	3311	3496	3762

Table 13.2.2B Geometry Factor, C_Δ, for Timber Rivet Connections Loaded Perpendicular to Grain

$\dfrac{e_p}{(n_c-1)S_q}$	C_Δ	$\dfrac{e_p}{(n_c-1)S_q}$	C_Δ
0.1	5.76	3.2	0.79
0.2	3.19	3.6	0.77
0.3	2.36	4.0	0.76
0.4	2.00	5.0	0.72
0.5	1.77	6.0	0.70
0.6	1.61	7.0	0.68
0.7	1.47	8.0	0.66
0.8	1.36	9.0	0.64
0.9	1.28	10.0	0.63
1.0	1.20	12.0	0.61
1.2	1.10	14.0	0.59
1.4	1.02	16.0	0.57
1.6	0.96	18.0	0.56
1.8	0.92	20.0	0.55
2.0	0.89	25.0	0.53
2.4	0.85	30.0	0.51
2.8	0.81		

TIMBER RIVETS

13

SHEAR WALLS AND DIAPHRAGMS

14

14.1 General

14.1.1 Application

Chapter 14 applies to the design of wood structural panel and lumber sheathed shear walls and diaphragms acting as elements of the lateral force-resisting system.

14.1.2 Definitions

14.1.2.1 The term "diaphragm" refers to a roof, floor, or other membrane or bracing system acting to transfer lateral forces to the vertical resisting elements.

14.1.2.2 The term "shear wall" refers to a wall designed to resist lateral forces parallel to the plane of the wall (sometimes referred to as a vertical diaphragm).

14.1.3 Framing Members

All framing including boundary members provided at shear wall and diaphragm perimeters, openings, and discontinuities and re-entrant corners shall be proportioned to resist the induced forces.

14.1.4 Fasteners

Values of fastener strength shall be determined in accordance with provisions of this standard.

14.1.5 Sheathing

Sheathing shall be proportioned to resist induced forces. The resistance of wood structural panel sheathing and lumber sheathing shall be investigated in accordance with the provisions of this standard.

14.2 Design Principles

Shear walls and diaphragms shall be designed according to a beam analogy with sheathing resisting in plane shear and framing members resisting axial forces or, alternate methods based on rational analysis. Design shall include consideration of sheathing, framing, fasteners, boundary members, and all required connections.

14.3 Shear Walls

14.3.1 Definitions

14.3.1.1 The term "shear wall height" refers to:
(1) The maximum clear height from the top of the foundation to the bottom of diaphragm framing above or,
(2) the maximum clear height from the top of diaphragm below to the bottom of diaphragm framing above. Where the diaphragm framing is sloped, the average height to the diaphragm framing above may be used.

14.3.1.2 The term "shear wall width" refers to the dimension of a shear wall in the direction of application of force and is measured as the dimension between boundary elements of the shear wall (in many cases, this will match the sheathed dimension).

14.3.1.3 The term "shear wall aspect ratio" refers to the ratio of height-to-width of a shear wall.

14.3.2 Shear Wall Anchorage

Connections shall be provided between the shear wall and attached components to transmit the induced forces.

14.3.3 Shear Force

The design shear force per unit length shall not exceed the adjusted shear wall shear resistance per unit length, $D \leq D'$.

14.3.4 Shear Resistance

The adjusted shear resistance, D', shall be determined by using principles of engineering mechanics using values of fastener strength and sheathing through-the-thickness shear resistance or, alternatively, from approved tables.

14.4 Diaphragms

14.4.1 Definitions

14.4.1.1 The term "collector" refers to a diaphragm element parallel and in line with the applied force that collects and transfers diaphragm shear forces to the vertical elements of the lateral-force-resisting system and/or distributes forces within the diaphragm.

14.4.1.2 The term "diaphragm chord" refers to a diaphragm boundary element perpendicular to the applied load that is assumed to take axial stresses due to the diaphragm moment.

14.4.1.3 The term "diaphragm length" (see Figure 14A) refers to the dimension of a diaphragm in the direction perpendicular to the application of force and is measured as the distance between vertical elements of the lateral-force-resisting system (in many cases, this will match the sheathed dimensions).

14.4.1.4 The term "diaphragm width" refers to the dimension of a diaphragm in the direction of application of force and is measured as the distance between diaphragm chords (in many cases, this will match the sheathed dimensions).

14.3.5 Shear Wall Deflection

When required in the design, the deflection of a shear wall shall be calculated in accordance with principles of engineering mechanics or by other approved methods.

14.4.1.5 The term "diaphragm aspect ratio" refers to the ratio of length to width of a diaphragm.

14.4.2 Shear Force

The design shear force per unit length shall not exceed the adjusted diaphragm shear resistance per unit length, D ≤ D'.

14.4.3 Shear Resistance

The adjusted shear resistance, D', shall be determined by using principles of engineering mechanics using values of fastener strength and sheathing through-the-thickness shear resistance or, alternatively, from approved tables.

14.4.4 Diaphragm Deflection

When required in the design, the deflection of a diaphragm shall be calculated in accordance with principles of engineering mechanics or by other approved methods.

SHEAR WALLS ANS DIAPHRAGMS

14

Figure 14A   Diaphragm Length and Width

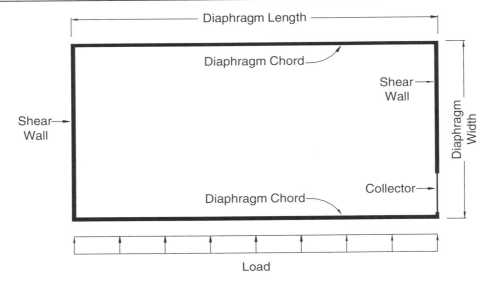

SPECIAL LOADING CONDITIONS

15

15.1 Lateral Distribution of a Concentrated Load

15.1.1 Lateral Distribution of a Concentrated Load for Moment

When a concentrated load at the center of the beam span is distributed to adjacent parallel beams by a wood or concrete-slab floor, the load on the beam nearest the point of application shall be determined by multiplying the load by the following factors:

Table 15.1.1   Lateral Distribution Factors for Moment

Kind of Floor	Load on Critical Beam (for one traffic lane[2])
2" plank	$S/4.0$[1]
4" nail laminated	$S/4.5$[1]
6" nail laminated	$S/5.0$[1]
Concrete, structurally designed	$S/6.0$[1]

1. S = average spacing of beams, ft. If S exceeds the denominator of the factor, the load on the two adjacent beams shall be the reactions of the load, with the assumption that the floor slab between the beams acts as a simple beam.
2. See Reference 48 for additional information concerning two or more traffic lanes.

15.1.2 Lateral Distribution of a Concentrated Load for Shear

When the load distribution for moment at the center of a beam is known or assumed to correspond to specific values in the first two columns of Table 15.1.2, the distribution to adjacent parallel beams when loaded at or near the quarter point (the approximate point of maximum shear) shall be assumed to be the corresponding values in the last two columns of Table 15.1.2.

Table 15.1.2   Lateral Distribution in Terms of Proportion of Total Load

Load Applied at Center of Span		Load Applied at 1/4 Point of Span	
Center Beam	Distribution to Side Beams	Center Beam	Distribution to Side Beams
1.00	0	1.00	0
0.90	0.10	0.94	0.06
0.80	0.20	0.87	0.13
0.70	0.30	0.79	0.21
0.60	0.40	0.69	0.31
0.50	0.50	0.58	0.42
0.40	0.60	0.44	0.56
0.33	0.67	0.33	0.67

15.2 Spaced Columns

15.2.1 General

15.2.1.1 The design load for a spaced column shall be the sum of the design loads for each of its individual members.

15.2.1.2 The increased load capacity of a spaced column due to the end-fixity developed by the split ring or shear plate connectors and end blocks is effective only in the direction perpendicular to the wide faces of the individual members (direction parallel to dimension d_1, in Figure 15A). The capacity of a spaced column in the direction parallel to the wide faces of the individual members (direction parallel to dimension d_2 in Figure 15A) shall be subject to the provisions for simple solid columns, as set forth in 15.2.3.

Figure 15A　Spaced Column Joined by Split Ring or Shear Plate Connectors

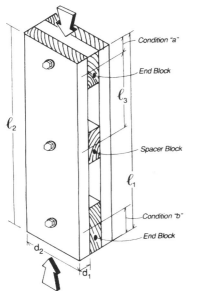

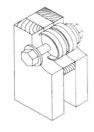

Typical shear plate connection in end block of spaced column

Spaced column

Condition "a": end distance $\leq \ell_1/20$

ℓ_1 and ℓ_2 = distances between points of lateral support in planes 1 and 2, measured from center to center of lateral supports for continuous spaced columns, and measured from end to end for simple spaced columns, inches.

ℓ_3 = Distance from center of spacer block to centroid of the group of split ring or shear plate connectors in end blocks, inches.

d_1 and d_2 = cross-sectional dimensions of individual rectangular compression members in planes of lateral support, inches.

Condition "b": $\ell_1/20$ < end distance $\leq \ell_1/10$

15.2.2 Spacer and End Block Provisions

15.2.2.1 Spaced columns shall be classified as to end fixity either as condition "a" or condition "b" (see Figure 15A), as follows:

(a) For condition "a", the centroid of the split ring or shear plate connector, or the group of connectors, in the end block shall be within $\ell_1/20$ from the column end.

(b) For condition "b", the centroid of the split ring or shear plate connector, or the group of connectors, in the end block shall be between $\ell_1/20$ and $\ell_1/10$ from the column end.

15.2.2.2 When a single spacer block is located within the middle 1/10 of the column length, ℓ_1, split ring or shear plate connectors shall not be required for this block. If there are two or more spacer blocks, split ring or shear plate connectors shall be required and the distance between two adjacent blocks shall not exceed ½ the distance between centers of split ring or shear plate connectors in the end blocks.

15.2.2.3 For spaced columns used as compression members of a truss, a panel point which is stayed laterally shall be considered as the end of the spaced column, and the portion of the web members, between the individual pieces making up a spaced column, shall be permitted to be considered as the end blocks.

15.2.2.4 Thickness of spacer and end blocks shall not be less than that of individual members of the spaced column nor shall thickness, width, and length of spacer and end blocks be less than required for split ring or shear plate connectors of a size and number capable of carrying the load computed in 15.2.2.5.

15.2.2.5 To obtain spaced column action the split ring or shear plate connectors in each mutually contacting surface of end block and individual member at each end of a spaced column shall be of a size and number to provide a load capacity in pounds equal to the required cross-sectional area in square inches of one of the individual members times the appropriate end spacer block constant, K_S, determined from the following equations:

Species Group	End Spacer Block Constant, K_S
A	$K_S = 9.55\,(\ell_1/d_1 - 11) \leq 468$
B	$K_S = 8.14\,(\ell_1/d_1 - 11) \leq 399$
C	$K_S = 6.73\,(\ell_1/d_1 - 11) \leq 330$
D	$K_S = 5.32\,(\ell_1/d_1 - 11) \leq 261$

If spaced columns are a part of a truss system or other similar framing, the split ring or shear plate connectors required by the connection provisions in Chapter 12 of this Specification shall be checked against the end spacer block constants, K_S, specified above.

15.2.3 Column Stability Factor, C_P

15.2.3.1 The effective column length, ℓ_e, for a spaced column shall be determined in accordance with principles of engineering mechanics. One method for determining effective column length, when end-fixity conditions are known, is to multiply actual column length by the appropriate effective length factor specified in Appendix G, $\ell_e = (K_e)(\ell)$, except that the effective column length, ℓ_e, shall not be less than the actual column length, ℓ.

SPECIAL LOADING CONDITIONS

15

15.2.3.2 For individual members of a spaced column (see Figure 15A):

(a) ℓ_1/d_1 shall not exceed 80, where ℓ_1 is the distance between lateral supports that provide restraint perpendicular to the wide faces of the individual members.

(b) ℓ_2/d_2 shall not exceed 50, where ℓ_2 is the distance between lateral supports that provide restraint in a direction parallel to the wide faces of the individual members.

(c) ℓ_3/d_1 shall not exceed 40, where ℓ_3 is the distance between the center of the spacer block and the centroid of the group of split ring or shear plate connectors in an end block.

15.2.3.3 The column stability factor shall be calculated as follows:

$$C_P = \frac{1+\left(F_{cE}/F_c^*\right)}{2c} - \sqrt{\left[\frac{1+\left(F_{cE}/F_c^*\right)}{2c}\right]^2 - \frac{F_{cE}/F_c^*}{c}} \quad (15.2\text{-}1)$$

where:

F_c^* = reference compression design value parallel to grain multiplied by all applicable adjustment factors except C_P (see 2.3)

$$F_{cE} = \frac{0.822\,K_x\,E_{min}'}{\left(\ell_e/d\right)^2}$$

K_x = 2.5 for fixity condition "a"

= 3.0 for fixity condition "b"

c = 0.8 for sawn lumber

= 0.9 for structural glued laminated timber or structural composite lumber

15.2.3.4 When individual members of a spaced column are of different species, grades, or thicknesses, the lesser adjusted compression parallel to grain design value, F_c', for the weaker member shall apply to both members.

15.2.3.5 The adjusted compression parallel to grain design value, F_c', for a spaced column shall not exceed the adjusted compression parallel to grain design value, F_c', for the individual members evaluated as solid columns without regard to fixity in accordance with 3.7 using the column slenderness ratio ℓ_2/d_2 (see Figure 15A).

15.2.3.6 For especially severe service conditions and/or extraordinary hazard, use of lower adjusted design values may be necessary. See Appendix H for background information concerning column stability calculations and Appendix F for information concerning coefficient of variation in modulus of elasticity (COV_E).

15.2.3.7 The equations in 3.9 for combined flexure and axial loading apply to spaced columns only for uniaxial bending in a direction parallel to the wide face of the individual member (dimension d_2 in Figure 15A).

15.3 Built-Up Columns

15.3.1 General

The following provisions apply to nailed or bolted built-up columns with 2 to 5 laminations in which:

(a) each lamination has a rectangular cross section and is at least 1-1/2" thick, $t \geq 1\text{-}1/2$".

(b) all laminations have the same depth (face width), d

(c) faces of adjacent laminations are in contact.

(d) all laminations are full column length.

(e) the connection requirements in 15.3.3 or 15.3.4 are met.

Nailed or bolted built-up columns not meeting the preceding limitations shall have individual laminations designed in accordance with 3.6.3 and 3.7. When individual laminations are of different species, grades, or thicknesses, the lesser adjusted compression parallel to grain design value, F_c', and modulus of elasticity for beam and column stability, E_{min}', for the weakest lamination shall apply.

15.3.2 Column Stability Factor, C_P

15.3.2.1 The effective column length, ℓ_e, for a built-up column shall be determined in accordance with principles of engineering mechanics. One method for determining effective column length, when end-fixity conditions are known, is to multiply actual column length by the appropriate effective length factor specified in Appendix G, $\ell_e = (K_e)(\ell)$.

15.3.2.2 The slenderness ratios ℓ_{e1}/d_1 and ℓ_{e2}/d_2 (see Figure 15B) where each ratio has been adjusted by the appropriate buckling length coefficient, K_e, from Appendix G, shall be determined. Each ratio shall be used to calculate a column stability factor, C_P, per section 15.3.2.4 and the smaller C_P shall be used in determining the adjusted compression design value parallel to grain, F_c', for the column. F_c' for built-up columns need not be less than F_c' for the individual laminations designed as individual solid columns per section 3.7.

15.3.2.3 The slenderness ratio, ℓ_e/d, for built-up columns shall not exceed 50, except that during construction ℓ_e/d shall not exceed 75.

15.3.2.4 The column stability factor shall be calculated as follows:

$$C_P = K_f \left[\frac{1+\left(F_{cE}/F_c^*\right)}{2c} - \sqrt{\left[\frac{1+\left(F_{cE}/F_c^*\right)}{2c}\right]^2 - \frac{F_{cE}/F_c^*}{c}} \right] \qquad (15.3-1)$$

where:

F_c^* = reference compression design value parallel to grain multiplied by all applicable modification factors except C_p (see 2.3)

$F_{cE} = \dfrac{0.822\,E_{min}'}{\left(\ell_e/d\right)^2}$

K_f = 0.6 for built-up columns where ℓ_{e2}/d_2 is used to calculate F_{cE} and the built-up columns are nailed in accordance with 15.3.3

K_f = 0.75 for built-up columns where ℓ_{e2}/d_2 is used to calculate F_{cE} and the built-up columns are bolted in accordance with 15.3.4

K_f = 1.0 for built-up columns where ℓ_{e1}/d_1 is used to calculate F_{cE} and the built-up columns are either nailed or bolted in accordance with 15.3.3 or 15.3.4, respectively

c = 0.8 for sawn lumber

c = 0.9 for structural glued laminated timber or structural composite lumber

15.3.2.5 For especially severe service conditions and/or extraordinary hazard, use of lower adjusted design values may be necessary. See Appendix H for background information concerning column stability calculations and Appendix F for information concerning coefficient of variation in modulus of elasticity (COV_E).

Figure 15B   Mechanically Laminated Built-Up Columns

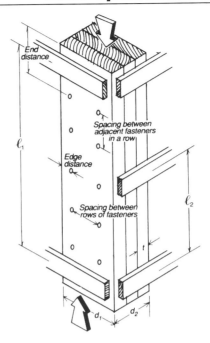

15.3.3 Nailed Built-Up Columns

15.3.3.1 The provisions in 15.3.1 and 15.3.2 apply to nailed built-up columns (see Figure 15C) in which:

(a) adjacent nails are driven from opposite sides of the column
(b) all nails penetrate at least 3/4 of the thickness of the last lamination
(c) $15D \le$ end distance $\le 18D$
(d) $20D \le$ spacing between adjacent nails in a row $\le 6t_{min}$
(e) $10D \le$ spacing between rows of nails $\le 20D$
(f) $5D \le$ edge distance $\le 20D$
(g) 2 or more longitudinal rows of nails are provided when $d > 3t_{min}$

where:

D = nail diameter

d = depth (face width) of individual lamination

t_{min} = thickness of thinnest lamination

When only one longitudinal row of nails is required, adjacent nails shall be staggered (see Figure 15C). When three or more longitudinal rows of nails are used, nails in adjacent rows shall be staggered.

SPECIAL LOADING CONDITIONS

15

Figure 15C   Typical Nailing Schedules for Built-Up Columns

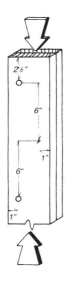

Two 2"x 4" laminations with one row of staggered 10d common wire nails (D = 0.148", L = 3")

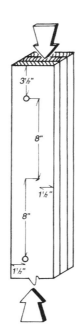

Three 2"x 4" laminations with one row of staggered 30d common wire nails (D = 0.207", L = 4-1/2")

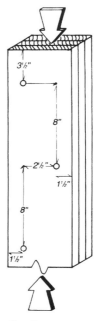

Three 2"x 6" laminations with two rows of 30d common wire nails (D = 0.207", L = 4-1/2")

15.3.4 Bolted Built-Up Columns

15.3.4.1 The provisions in 15.3.1 and 15.3.2 apply to bolted built-up columns in which:

(a) a metal plate or washer is provided between the wood and the bolt head, and between the wood and the nut

(b) nuts are tightened to insure that faces of adjacent laminations are in contact

(c) for softwoods: $7D \leq$ end distance $\leq 8.4D$
    for hardwoods: $5D \leq$ end distance $\leq 6D$

(d) $4D \leq$ spacing between adjacent bolts in a row $\leq 6t_{min}$

(e) $1.5D \leq$ spacing between rows of bolts $\leq 10D$

(f) $1.5D \leq$ edge distance $\leq 10D$

(g) 2 or more longitudinal rows of bolts are provided when $d > 3t_{min}$

where:

D = bolt diameter

d = depth (face width) of individual lamination

t_{min} = thickness of thinnest lamination

15.3.4.2 Figure 15D provides an example of a bolting schedule which meets the preceding connection requirements.

Figure 15D Typical Bolting Schedules for Built-Up Columns

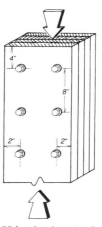

Four 2" x 8" laminations (softwoods) with two rows of ½" diameter bolts.

15.4 Wood Columns with Side Loads and Eccentricity

15.4.1 General Equations

One design method that allows calculation of the direct compression load that an eccentrically loaded column, or one with a side load, is capable of sustaining is as follows:

(a) Members subjected to a combination of bending from eccentricity and/or side loads about one or both principal axes, and axial compression, shall be proportioned so that:

$$\left(\frac{f_c}{F_c'}\right)^2 + \frac{f_{b1} + f_c(6e_1/d_1)[1+0.234(f_c/F_{cE1})]}{F_{b1}'\left[1-(f_c/F_{cE1})\right]} + \qquad (15.4\text{-}1)$$

$$\frac{f_{b2} + f_c(6e_2/d_2)\left\{1+0.234(f_c/F_{cE2})+0.234\left[\dfrac{f_{b1}+f_c(6e_1/d_1)}{F_{bE}}\right]^2\right\}}{F_{b2}'\left\{1-(f_c/F_{cE2})-\left[\dfrac{f_{b1}+f_c(6e_1/d_1)}{F_{bE}}\right]^2\right\}} \le 1.0$$

(b) Members subjected to a combination of bending and compression from an eccentric axial load about one or both principal axes, shall be proportioned so that:

$$\left(\frac{f_c}{F_c'}\right)^2 + \frac{f_c(6e_1/d_1)[1+0.234(f_c/F_{cE1})]}{F_{b1}'\left[1-(f_c/F_{cE1})\right]} + \qquad (15.4\text{-}2)$$

$$\frac{f_c(6e_2/d_2)\left\{1+0.234(f_c/F_{cE2})+0.234\left[\dfrac{f_c(6e_1/d_1)}{F_{bE}}\right]^2\right\}}{F_{b2}'\left\{1-(f_c/F_{cE2})-\left[\dfrac{f_c(6e_1/d_1)}{F_{bE}}\right]^2\right\}} \le 1.0$$

where:

$$f_c < F_{cE1} = \frac{0.822\,E_{min}'}{\left(\ell_{e1}/d_1\right)^2} \quad \begin{array}{l}\text{for either uniaxial edgewise}\\ \text{bending or biaxial bending}\end{array}$$

and

$$f_c < F_{cE2} = \frac{0.822\,E_{min}'}{\left(\ell_{e2}/d_2\right)^2} \quad \begin{array}{l}\text{for uniaxial flatwise bending}\\ \text{or biaxial bending}\end{array}$$

and

$$f_{b1} < F_{bE} = \frac{1.20\,E_{min}'}{R_B^2} \quad \text{for biaxial bending}$$

f_c = compression stress parallel to grain due to axial load

f_{b1} = edgewise bending stress due to side loads on narrow face only

f_{b2} = flatwise bending stress due to side loads on wide face only

F_c' = adjusted compression design value parallel to grain that would be permitted if axial compressive stress only existed, determined in accordance with 2.3 and 3.7

F_{b1}' = adjusted edgewise bending design value that would be permitted if edgewise bending stress only existed, determined in accordance with 2.3 and 3.3.3

F_{b2}' = adjusted flatwise bending design value that would be permitted if flatwise bending stress only existed, determined in accordance with 2.3 and 3.3.3

R_B = slenderness ratio of bending member (see 3.3.3)

d_1 = wide face dimension

d_2 = narrow face dimension

e_1 = eccentricity, measured parallel to wide face from centerline of column to centerline of axial load

e_2 = eccentricity, measured parallel to narrow face from centerline of column to centerline of axial load

Effective column lengths, ℓ_{e1} and ℓ_{e2}, shall be determined in accordance with 3.7.1.2. F_{cE1} and F_{cE2} shall be determined in accordance with 3.7. F_{bE} shall be determined in accordance with 3.3.3.

SPECIAL LOADING CONDITIONS

15

15.4.2 Columns with Side Brackets

15.4.2.1 The formulas in 15.4.1 assume that the eccentric load is applied at the end of the column. One design method that allows calculation of the actual bending stress, f_b, if the eccentric load is applied by a bracket within the upper quarter of the length of the column is as follows.

5.4.2.2 Assume that a bracket load, P, at a distance, a, from the center of the column (Figure 15E), is replaced by the same load, P, centrally applied at the top of the column, plus a side load, P_s, applied at midheight. Calculate P_s from the following formula:

$$P_s = \frac{3P\,a\,\ell_p}{\ell^2} \qquad\qquad (15.4\text{-}3)$$

where:

> P = actual load on bracket, lbs.
>
> P_s = assumed horizontal side load placed at center of height of column, lbs.
>
> a = horizontal distance from load on bracket to center of column, in.
>
> ℓ = total length of column, in.
>
> ℓ_p = distance measured vertically from point of application of load on bracket to farther end of column, in.

The assumed centrally applied load, P, shall be added to other concentric column loads, and the calculated side load, P_s, shall be used to determine the actual bending stress, f_b, for use in the formula for concentric end and side loading.

Figure 15E   Eccentrically Loaded Column

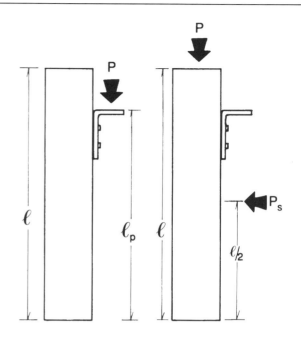

FIRE DESIGN OF WOOD MEMBERS

16

16.1 General

Chapter 16 establishes general fire design provisions that apply to all wood structural members and connections covered under this Specification, unless otherwise noted. Each wood member or connection shall be of sufficient size and capacity to carry the applied loads without exceeding the design provisions specified herein. Reference design values and specific design provisions applicable to particular wood products or connections to be used with the provisions of this Chapter are given in other Chapters of this Specification.

16.2 Design Procedures for Exposed Wood Members

The induced stress shall not exceed the resisting strength which have been adjusted for fire exposure. Wood member design provisions herein are limited to fire resistance calculations not exceeding 2 hours.

16.2.1 Char Rate

The effective char rate to be used in the this procedure can be estimated from published nominal 1-hour char rate data using the following equation:

$$\beta_{eff} = \frac{1.2\beta_n}{t^{0.187}}$$
(16.2-1)

where:

β_{eff} = effective char rate (in./hr.), adjusted for exposure time, t

β_n = nominal char rate (in./hr.), linear char rate based on 1-hour exposure

t = exposure time (hrs.)

A nominal char rate, β_n, of 1.5 in./hr. is commonly assumed for solid sawn and structural glued laminated softwood members. For β_n = 1.5 in./hr., the effective char rates, β_{eff}, and effective char layer thicknesses, a_{char}, for each exposed surface are shown in Table 16.2.1.

Table 16.2.1 Effective Char Rates and Char Layer Thicknesses (for β_n = 1.5 in./hr.)

Required Fire Endurance (hr.)	Effective Char Rate, β_{eff} (in./hr.)	Effective Char Layer Thickness, a_{char} (in.)
1-Hour	1.8	1.8
1½-Hour	1.67	2.5
2-Hour	1.58	3.2

Section properties shall be calculated using standard equations for area, section modulus, and moment of inertia using the reduced cross-sectional dimensions. The dimensions are reduced by the effective char layer thickness, a_{char}, for each surface exposed to fire.

16.2.2 Member Strength

For solid sawn, structural glued laminated timber, and structural composite lumber wood members, the average member strength can be approximated by multiplying reference design values (F_b, F_t, F_c, F_{bE}, F_{cE}) by the adjustment factors specified in Table 16.2.2.

All member strength and cross-sectional properties shall be adjusted prior to use of the interaction calculations in 3.9 or 15.4.

16.2.3 Design of Members

The induced stress calculated using reduced section properties determined in 16.2.1 shall not exceed the member strength determined in 16.2.2.

16.2.4 Special Provisions for Structural Glued Laminated Timber Beams

For structural glued laminated timber bending members given in Table 5A and rated for 1-hour fire endurance, an outer tension lamination shall be substituted for a core lamination on the tension side for unbalanced beams and on both sides for balanced beams. For structural glued laminated timber bending members given in Table 5A and rated for 1½- or 2-hour fire endurance, 2 outer tension laminations shall be substituted for 2 core laminations on the tension side for unbalanced beams and on both sides for balanced beams.

16.2.5 Provisions for Timber Decks

Timber decks consist of planks that are at least 2" (actual) thick. The planks shall span the distance between supporting beams. Single and double tongue-and-groove (T&G) decking shall be designed as an assembly of wood beams fully exposed on one face. Butt-jointed decking shall be designed as an assembly of wood beams partially exposed on the sides and fully exposed on one face. To compute the effects of partial exposure of the decking on its sides, the char rate for this limited exposure shall be reduced to 33% of the effective char rate. These calculation procedures do not address thermal separation.

Table 16.2.2 Adjustment Factors for Fire Design[1]

			ASD					
			Design Stress to Member Strength Factor	Size Factor[2]	Volume Factor[2]	Flat Use Factor[2]	Beam Stability Factor[3]	Column Stability Factor[3]
Bending Strength	F_b	x	2.85	C_F	C_V	C_{fu}	C_L	-
Tensile Strength	F_t	x	2.85	C_F	-	-	-	-
Compression Strength	F_c	x	2.58	C_F	-	-	-	C_P
Beam Buckling Strength	F_{bE}	x	2.03	-	-	-	-	-
Column Buckling Strength	F_{cE}	x	2.03	-	-	-	-	-

1. See 4.3, 5.3 and 8.3 for applicability of adjustment factors for specific products.
2. Factor shall be based on initial cross-section dimensions.
3. Factor shall be based on reduced cross-section dimensions.

16.3 Wood Connections

Where fire endurance is required, connectors and fasteners shall be protected from fire exposure by wood, fire-rated gypsum board, or any coating approved for the required endurance time.

A

APPENDIX

Appendix A (Non-mandatory) Construction and Design Practices

A.1 Care of Material

Lumber shall be so handled and covered as to prevent marring and moisture absorption from snow or rain.

A.2 Foundations

A.2.1 Foundations shall be adequate to support the building or structure and any required loads, without excessive or unequal settlement or uplift.

A.2.2 Good construction practices generally eliminate decay or termite damage. Such practices are designed to prevent conditions which would be conducive to decay and insect attack. The building site shall be graded to provide drainage away from the structure. All roots and scraps of lumber shall be removed from the immediate vicinity of the building before backfilling.

A.3 Structural Design

Consideration shall be given in design to the possible effect of cross-grain dimensional changes which may occur in lumber fabricated or erected in a green condition (i.e., provisions shall be made in the design so that if dimensional changes caused by seasoning to moisture equilibrium occur, the structure will move as a whole, and the differential movement of similar parts and members meeting at connections will be a minimum).

A.4 Drainage

In exterior structures, the design shall be such as to minimize pockets in which moisture can accumulate, or adequate caps, drainage, and drips shall be provided.

A.5 Camber

Adequate camber in trusses to give proper appearance and to counteract any deflection from loading should be provided. For timber connector construction, such camber shall be permitted to be estimated from the formula:

$$\Delta = \frac{K_1 L^3 + K_2 L^2}{H} \qquad (A-1)$$

where:

Δ = camber at center of truss, in.

L = truss span, ft

H = truss height at center, ft

K_1 = 0.000032 for any type of truss

K_2 = 0.0028 for flat and pitched trusses

K_2 = 0.00063 for bowstring trusses (i.e., trusses without splices in upper chord)

A.6 Erection

A.6.1 Provision shall be made to prevent the overstressing of members or connections during erection.

A.6.2 Bolted connections shall be snugly tightened, but not to the extent of crushing wood under washers.

A.6.3 Adequate bracing shall be provided until permanent bracing and/or diaphragms are installed.

A.7 Inspection

Provision should be made for competent inspection of materials and workmanship.

A.8 Maintenance

There shall be competent inspection and tightening of bolts in connections of trusses and structural frames.

A.9 Wood Column Bracing

In buildings, for forces acting in a direction parallel to the truss or beam, column bracing shall be permitted to be provided by knee braces or, in the case of trusses, by extending the column to the top chord of the truss where the bottom and top chords are separated sufficiently to provide adequate bracing action. In a direction perpendicular to the truss or beam, bracing shall be permitted to be provided by wall construction, knee braces, or bracing between columns. Such bracing between columns should be installed preferably in the same bays as the bracing between trusses.

A.10 Truss Bracing

In buildings, truss bracing to resist lateral forces shall be permitted as follows:

(a) Diagonal lateral bracing between top chords of trusses shall be permitted to be omitted when the provisions of Appendix A.11 are followed or when the roof joists rest on and are securely fastened to the top chords of the trusses and are covered with wood sheathing. Where sheathing other than wood is applied, top chord diagonal lateral bracing should be installed.

(b) In all cases, vertical sway bracing should be installed in each third or fourth bay at intervals of approximately 35 feet measured parallel to trusses. Also, bottom chord lateral bracing should be installed in the same bays as the vertical sway bracing, where practical, and should extend from side wall to side wall. In addition, struts should be installed between bottom chords at the same truss panels as vertical sway bracing and should extend continuously from end wall to end wall. If the roof construction does not provide proper top chord strut action, separate additional members should be provided.

A.11 Lateral Support of Arches, Compression Chords of Trusses and Studs

A.11.1 When roof joists or purlins are used between arches or compression chords, or when roof joists or purlins are placed on top of an arch or compression chord, and are securely fastened to the arch or compression chord, the largest value of ℓ_e/d, calculated using the depth of the arch or compression chord or calculated using the breadth (least dimension) of the arch or compression chord between points of intermittent lateral support, shall be used. The roof joists or purlins should be placed to account for shrinkage (for example by placing the upper edges of unseasoned joists approximately 5% of the joist depth above the tops of the arch or chord), but also placed low enough to provide adequate lateral support.

A.11.2 When planks are placed on top of an arch or compression chord, and securely fastened to the arch or compression chord, or when sheathing is nailed properly to the top chord of trussed rafters, the depth rather than the breadth of the arch, compression chord, or trussed rafter shall be permitted to be used as the least dimension in determining ℓ_e/d.

A.11.3 When stud walls in light frame construction are adequately sheathed on at least one side, the depth, rather than breadth of the stud, shall be permitted to be taken as the least dimension in calculating the ℓ_e/d ratio. The sheathing shall be shown by experience to provide lateral support and shall be adequately fastened.

Appendix B (Non-mandatory) Load Duration (ASD Only)

B.1 Adjustment of Reference Design Values for Load Duration

B.1.1 Normal Load Duration. The reference design values in this Specification are for normal load duration. Normal load duration contemplates fully stressing a member to its allowable design value by the application of the full design load for a cumulative duration of approximately 10 years and/or the application of 90% of the full design load continuously throughout the remainder of the life of the structure, without encroaching on the factor of safety.

B.1.2 Other Load Durations. Since tests have shown that wood has the property of carrying substantially greater maximum loads for short durations than for long durations of loading, reference design values for normal load duration shall be multiplied by load duration factors, C_D, for other durations of load (see Figure B1). Load duration factors do not apply to reference modulus of elasticity design values, E, nor to reference compression design values perpendicular to grain, $F_{c\perp}$, based on a deformation limit.

(a) When the member is fully stressed to the adjusted design value by application of the full design load permanently, or for a cumulative total of more than 10 years, reference design values for normal load duration (except E and $F_{c\perp}$ based on a deformation limit) shall be multiplied by the load duration factor, $C_D = 0.90$.

(b) Likewise, when the duration of the full design load does not exceed the following durations, reference design values for normal load duration (except E and $F_{c\perp}$ based on a deformation limit) shall be multiplied by the following load duration factors:

C_D	Load Duration
1.15	two months duration
1.25	seven days duration
1.6	ten minutes duration
2.0	impact

(c) The 2 month load duration factor, $C_D = 1.15$, is applicable to design snow loads based on ASCE 7. Other load duration factors shall be permitted to be used where such adjustments are referenced to the duration of the design snow load in the specific location being considered.

(d) The 10 minutes load duration factor, $C_D = 1.6$, is applicable to design earthquake loads and design wind loads based on ASCE 7.

(e) Load duration factors greater than 1.6 shall not apply to structural members pressure-treated with water-borne preservatives (see Reference 30), or fire retardant chemicals. The impact load duration factor shall not apply to connections.

B.2 Combinations of Loads of Different Durations

When loads of different durations are applied simultaneously to members which have full lateral support to prevent buckling, the design of structural members and connections shall be based on the critical load combination determined from the following procedures:

(a) Determine the magnitude of each load that will occur on a structural member and accumulate subtotals of combinations of these loads. Design loads established by applicable building codes and standards may include load combination factors to adjust for probability of simultaneous occurrence of various loads (see Appendix B.4). Such load combination factors should be included in the load combination subtotals.

(b) Divide each subtotal by the load duration factor, C_D, for the shortest duration load in the combination of loads under consideration.

Shortest Load Duration in the Combination of Loads	Load Duration Factor, C_D
Permanent	0.9
Normal	1.0
Two Months	1.15
Seven Days	1.25
Ten Minutes	1.6
Impact	2.0

(c) The largest value thus obtained indicates the critical load combination to be used in designing the structural member or connection.

EXAMPLE: Determine the critical load combination for a structural member subjected to the following loads:

D = dead load established by applicable building code or standard

L = live load established by applicable building code or standard

S = snow load established by applicable building code or standard

W = wind load established by applicable building code or standard

The actual stress due to any combination of the above loads shall be less than or equal to the adjusted design value modified by the load duration factor, C_D, for the shortest duration load in that combination of loads:

Actual stress due to	(C_D)	x (Design value)
D	≤ (0.9)	x (design value)
D+L	≤ (1.0)	x (design value)
D+W	≤ (1.6)	x (design value)
D+L+S	≤ (1.15)	x (design value)
D+L+W	≤ (1.6)	x (design value)
D+S+W	≤ (1.6)	x (design value)
D+L+S+W	≤ (1.6)	x (design value)

The equations above may be specified by the applicable building code and shall be checked as required. Load combination factors specified by the applicable building code or standard should be included in the above equations, as specified in B.2(a).

B.3 Mechanical Connections

Load duration factors, C_D ≤ 1.6, apply to reference design values for connections, except when connection capacity is based on design of metal parts (see 10.2.3).

B.4 Load Combination Reduction Factors

Reductions in total design load for certain combinations of loads account for the reduced probability of simultaneous occurrence of the various design loads. Load duration factors, C_D, account for the relationship between wood strength and time under load. Load duration factors, C_D, are independent of load combination reduction factors, and both may be used in design calculations (see 1.4.4).

Figure B1 Load Duration Factors, C_D, for Various Load Durations

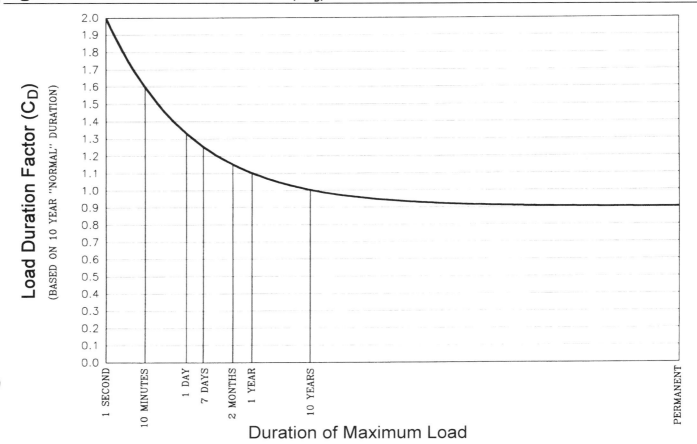

Appendix C (Non-mandatory) Temperature Effects

C.1

As wood is cooled below normal temperatures, its strength increases. When heated, its strength decreases. This temperature effect is immediate and its magnitude varies depending on the moisture content of the wood. Up to 150°F, the immediate effect is reversible. The member will recover essentially all its strength when the temperature is reduced to normal. Prolonged heating to temperatures above 150°F can cause a permanent loss of strength.

C.2

In some regions, structural members are periodically exposed to fairly elevated temperatures. However, the normal accompanying relative humidity generally is very low and, as a result, wood moisture contents also are low. The immediate effect of the periodic exposure to the elevated temperatures is less pronounced because of this dryness. Also, independently of temperature changes, wood strength properties generally increase with a decrease in moisture content. In recognition of these offsetting factors, it is traditional practice to use the reference design values from this Specification for ordinary temperature fluctuations and occasional short-term heating to temperatures up to 150°F.

C.3

When wood structural members are heated to temperatures up to 150°F for extended periods of time, adjustment of the reference design values in this Specification may be necessary (see 2.3.3 and 10.3.4). See Reference 53 for additional information concerning the effect of temperature on wood strength.

Appendix D (Non-mandatory) Lateral Stability of Beams

D.1

Slenderness ratios and related equations for adjusting reference bending design values for lateral buckling in 3.3.3 are based on theoretical analyses and beam verification tests.

D.2

Treatment of lateral buckling in beams parallels that for columns given in 3.7.1 and Appendix H. Beam stability calculations are based on slenderness ratio, R_B, defined as:

$$R_B = \sqrt{\frac{\ell_e d}{b^2}} \tag{D-1}$$

with ℓ_e as specified in 3.3.3.

D.3

For beams with rectangular cross section where R_B does not exceed 50, adjusted bending design values are obtained by the equation (where $C_L \le C_V$):

$$F_b' = F_b^* \left[\frac{1 + \left(F_{bE}/F_b^*\right)}{1.9} - \sqrt{\left[\frac{1 + \left(F_{bE}/F_b^*\right)}{1.9}\right]^2 - \frac{F_{bE}/F_b^*}{0.95}} \right] \tag{D-2}$$

where:

$$F_{bE} = \frac{1.20\, E_{min}'}{R_B^2} \tag{D-3}$$

F_b^* = reference bending design value multiplied by all applicable adjustment factors except C_{fu}, C_V, and C_L (see 2.3)

D.4

Reference modulus of elasticity for beam and column stability, E_{min}, in Equation D-3 is based on the following equation:

$$E_{min} = E\,[1 - 1.645\,COV_E](1.03)/1.66 \tag{D-4}$$

where:

E = reference modulus of elasticity

1.03 = adjustment factor to convert E values to a pure bending basis except that the factor is 1.05 for structural glued laminated timber

1.66 = factor of safety

COV_E = coefficient of variation in modulus of elasticity (see Appendix F)

E_{min} represents an approximate 5% lower exclusion value on pure bending modulus of elasticity, plus a 1.66 factor of safety.

D.5

For products with less E variability than visually graded sawn lumber, higher critical buckling design values (F_{bE}) may be calculated. For a product having a lower coefficient of variation in modulus of elasticity, use of Equations D-3 and D-4 will provide a 1.66 factor of safety at the 5% lower exclusion value.

Appendix E (Non-mandatory) Local Stresses in Fastener Groups

E.1 General

Where a fastener group is composed of closely spaced fasteners loaded parallel to grain, the capacity of the fastener group may be limited by wood failure at the net section or tear-out around the fasteners caused by local stresses. One method to evaluate member strength for local stresses around fastener groups is outlined in the following procedures.

E.1.1 Reference design values for timber rivet connections in Chapter 13 account for local stress effects and do not require further modification by procedures outlined in this Appendix.

E.1.2 The capacity of connections with closely spaced, large diameter bolts has been shown to be limited by the capacity of the wood surrounding the connection. Connections with groups of smaller diameter fasteners, such as typical nailed connections in wood-frame construction, may not be limited by wood capacity.

E.2 Net Section Tension Capacity

The adjusted tension capacity is calculated in accordance with provisions of 3.1.2 and 3.8.1 as follows:

$$Z_{NT}' = F_t' A_{net} \qquad \text{(E.2-1)}$$

where:

Z_{NT}' = adjusted tension capacity of net section area

F_t' = adjusted tension design value parallel to grain

A_{net} = net section area per 3.1.2

E.3 Row Tear-Out Capacity

The adjusted tear-out capacity of a row of fasteners can be estimated as follows:

$$Z_{RTi}' = n_i \frac{F_v' A_{critical}}{2} \qquad \text{(E.3-1)}$$

where:

Z_{RTi}' = adjusted row tear out capacity of row i

F_v' = adjusted shear design value parallel to grain

$A_{critical}$ = minimum shear area of any fastener in row i

n_i = number of fasteners in row i

E3.1 Assuming one shear line on each side of bolts in a row (observed in tests of bolted connections), Equation E.3-1 becomes:

$$Z_{RTi}' = \frac{F_v' t}{2} [n_i s_{critical}](2 \text{ shear lines}) \qquad \text{(E.3-2)}$$

$$= n_i F_v' t s_{critical}$$

where:

$s_{critical}$ = minimum spacing in row i taken as the lesser of the end distance or the spacing between fasteners in row i

t = thickness of member

The total adjusted row tear-out capacity of multiple rows of fasteners can be estimated as:

$$Z_{RT}' = \sum_{i=1}^{n_{row}} Z_{RTi}' \qquad \text{(E.3-3)}$$

where:

Z_{RT}' = adjusted row tear out capacity of multiple rows

n_{row} = number of rows

E.3.2 In Equation E.3-1, it is assumed that the induced shear stress varies from a maximum value of $f_v = F_v'$ to a minimum value of $f_v = 0$ along each shear line between fasteners in a row and that the change in shear stress/strain is linear along each shear line. The resulting triangular stress distribution on each shear line between fasteners in a row establishes an apparent shear stress equal to half of the adjusted design shear stress, $F_v'/2$, as shown in Equation E.3-1. This assumption is combined with the critical area concept for evaluating stresses in fastener groups and provides good agreement with results from tests of bolted connections.

E.3.3 Use of the minimum shear area of any fastener in a row for calculation of row tear-out capacity is based on the assumption that the smallest shear area between fasteners in a row will limit the capacity of the row of fasteners. Limited verification of this approach is provided from tests of bolted connections.

E.4 Group Tear-Out Capacity

The adjusted tear-out capacity of a group of "n" rows of fasteners can be estimated as:

$$Z_{GT}' = \frac{Z_{RT-1}'}{2} + \frac{Z_{RT-n}'}{2} + F_t' A_{group-net} \qquad \text{(E.4-1)}$$

where:

Z_{GT}' = adjusted group tear-out capacity

Z_{RT-1}' = adjusted row tear-out capacity of row 1 of fasteners bounding the critical group area

Z_{RT-n}' = adjusted row tear-out capacity of row n of fasteners bounding the critical group area

$A_{group-net}$ = critical group net section area between row 1 and row n

E.4.1 For groups of fasteners with non-uniform spacing between rows of fasteners various definitions of critical group area should be checked for group tear-out in combination with row tear-out to determine the adjusted capacity of the critical section.

E.5 Effects of Fastener Placement

E 5.1 Modification of fastener placement within a fastener group can be used to increase row tear-out and group tear-out capacity limited by local stresses around the fastener group. Increased spacing between fasteners in a row is one way to increase row tear-out capacity. Increased spacing between rows of fasteners is one way to increase group tear-out capacity.

E 5.2 Footnote 2 to Table 11.5.1D limits the spacing between outer rows of fasteners paralleling the member on a single splice plate to 5 inches. This requirement is imposed to limit local stresses resulting from shrinkage of wood members. When special detailing is used to address shrinkage, such as the use of slotted holes, the 5-inch limit can be adjusted.

E.6 Sample Solution of Staggered Bolts

Calculate the net section area tension, row tear-out, and group tear-out ASD adjusted design capacities for the double-shear bolted connection in Figure E1.

Main Member:
Combination 3 Douglas fir 3-1/8 x 12 glued laminated timber member
$F_t' = 1450$ psi
$F_v' = 240$ psi
Main member thickness, t_m: 3.125 inches
Main member width, w: 12 inches

Side Member:
A36 steel plates on each side
Side plate thickness, t_s: 0.25 inches

Connection Details:
Bolt diameter, D: 1 inch
Bolt hole diameter, D_h: 1.0625 inches
Adjusted ASD bolt design value, $Z_{\|}'$: 4380 lbs. (see NDS Table 11I. For this trial design, the group action factor, C_g, is taken as 1.0).
Spacing between rows: $s_{row} = 2.5D$

Adjusted ASD Connection Capacity, $nZ_{\|}'$:

$nZ_{\|}' = (8 \text{ bolts})(4,380 \text{ lbs.}) = 35,040 \text{ lbs.}$

Figure E1 Staggered Rows of Bolts

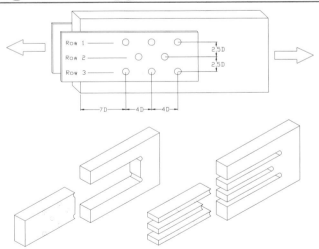

Adjusted ASD Net Section Area Tension Capacity, Z_{NT}':

$$Z_{NT}' = F_t' t \left[w - n_{row} D_h \right]$$

$$Z_{NT}' = (1,450 \text{ psi})(3.125")[12" - 3(1.0625")]$$
$$= 39,930 \text{ lbs.}$$

Adjusted ASD Row Tear-Out Capacity, Z_{RT}':

$$Z_{RTi}' = n_i F_v' t s_{critical}$$

$$
\begin{aligned}
Z_{RT-1}' &= 3(240 \text{ psi})(3.125")(4") = 9{,}000 \text{ lbs.} \\
Z_{RT-2}' &= 2(240 \text{ psi})(3.125")(4") = 6{,}000 \text{ lbs.} \\
Z_{RT-3}' &= 3(240 \text{ psi})(3.125")(4") = 9{,}000 \text{ lbs.}
\end{aligned}
$$

$$Z_{RT}' = \sum_{i=1}^{n_{row}} Z_{RTi}' = 9{,}000 + 6{,}000 + 9{,}000 = 24{,}000 \text{ lbs.}$$

Adjusted ASD Group Tear-Out Capacity, Z_{GT}':

$$Z_{GT}' = \frac{Z_{RT-1}'}{2} + \frac{Z_{RT-3}'}{2} + F_t' t \left[(n_{row} - 1)(s_{row} - D_h) \right]$$

$$
\begin{aligned}
Z_{GT}' &= (9{,}000 \text{ lbs.})/2 + (9{,}000 \text{ lbs.})/2 + \\
&\quad (1{,}450 \text{ psi})(3.125")[(3-1)(2.5" - 1.0625")] \\
&= 22{,}030 \text{ lbs.}
\end{aligned}
$$

In this sample calculation, the adjusted ASD connection capacity is limited to 22,030 pounds by group tear-out, Z_{GT}'.

E.7 Sample Solution of Row of Bolts

Calculate the net section area tension and row tear-out adjusted ASD design capacities for the single-shear single-row bolted connection represented in Figure E2.

Main and Side Members:
#2 grade Hem-Fir 2x4 lumber
$F_t' = 788$ psi
$F_v' = 145$ psi
Main member thickness, t_m: 3.5 inches
Side member thickness, t_s: 1.5 inches
Main and side member width, w: 3.5 inches

Connection Details:
Bolt diameter, D: 1/2 inch
Bolt hole diameter, D_h: 0.5625 inches
Adjusted ASD bolt design value, $Z_\parallel'$: 550 lbs. (See NDS Table 11A. For this trial design, the group action factor, C_g, is taken as 1.0).

Adjusted ASD Connection Capacity, $nZ_\parallel'$:

$$nZ_\parallel' = (3 \text{ bolts})(550 \text{ lbs.}) = 1{,}650 \text{ lbs.}$$

Adjusted ASD Net Section Area Tension Capacity, Z_{NT}':

$$Z_{NT}' = F_t' t \left[w - n_{row} D_h \right]$$

$$Z_{NT}' = (788 \text{ psi})(1.5")[3.5" - 1(0.5625")] = 3{,}470 \text{ lbs.}$$

Figure E2 Single Row of Bolts

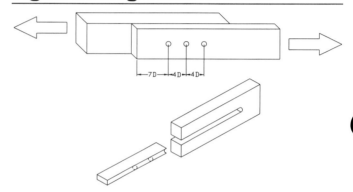

Adjusted ASD Row Tear-Out Capacity, Z_{RT}':

$$Z_{RTi}' = n_i F_v' t s_{critical}$$

$$Z_{RT1}' = 3(145 \text{ psi})(1.5")(2") = 1{,}310 \text{ lbs.}$$

In this sample calculation, the adjusted ASD connection capacity is limited to 1,310 pounds by row tear-out, Z_{RT}'.

E.8 Sample Solution of Row of Split Rings

Calculate the net section area tension and row tear-out adjusted ASD design capacities for the single-shear single-row split ring connection represented in Figure E3.

Main and Side Members:
#2 grade Southern Pine 2x4 lumber
$F_t' = 825$ psi
$F_v' = 175$ psi
Main member thickness, t_m: 1.5 inches
Side member thickness, t_s: 1.5 inches
Main and side member width, w: 3.5 inches

Connection Details:
Split ring diameter, D: 2.5 inches (see Appendix K for connector dimensions)
Adjusted ASD split ring design value, P': 2,730 lbs. (see NDS Table 12.2A. For this trial design, the group action factor, C_g, is taken as 1.0).

Adjusted ASD Connection Capacity, nP':

$$nP' = (2 \text{ split rings})(2,730 \text{ lbs.}) = 5,460 \text{ lbs.}$$

Adjusted ASD Net Section Area Tension Capacity, Z_{NT}':

$$Z_{NT}' = F_t' A_{net}$$

$$Z_{NT}' = F_t' [A_{2x4} - A_{bolt\text{-}hole} - A_{split\ ring\ projected\ area}]$$

$$Z_{NT}' = (825 \text{ psi})[5.25 \text{ in.}^2 - 1.5" (0.5625") - 1.1 \text{ in.}^2]$$
$$= 2,728 \text{ lbs.}$$

Figure E3   Single Row of Split Ring Connectors

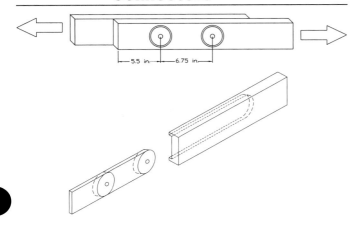

5.5 in.   6.75 in.

Adjusted ASD Row Tear-Out Capacity, Z_{RT}':

$$Z_{RTi}' = n_i \frac{F_v' A_{critical}}{2}$$

$$Z_{RT1}' = [(2 \text{ connectors})(175 \text{ psi})/2](21.735 \text{ in.}^2)$$
$$= 3,804 \text{ lbs.}$$

where:

$$A_{critical} = 21.735 \text{ in.}^2 \quad (\text{See Figure E4})$$

In this sample calculation, the adjusted ASD connection capacity is limited to 2,728 pounds by net section area tension capacity, Z_{NT}'.

Figure E4 $A_{critical}$ for Split Ring Connection

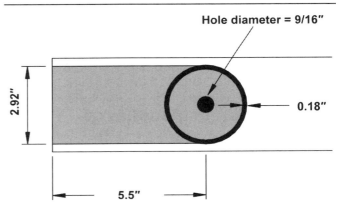

Hole diameter = 9/16"

2.92"

0.18"

5.5"

$$A_{critical} = A_{edge\ plane} + A_{bottom\ plane\ net}$$
$$= 21.735 \text{ in.}^2$$

$$A_{edge\ plane} = (2 \text{ shear lines}) (\text{groove depth})(s_{critical})$$
$$= (2 \text{ shear lines}) (0.375")(5.5") = 4.125 \text{ in.}^2$$

$$A_{bottom\ plane\ net} = (A_{bottom\ plane}) - (A_{split\ ring\ groove}) - (A_{bolt\ hole})$$
$$= [(5.5")(2.92") + (\pi)(2.92")^2/8] - (\pi/4)[(2.92")^2 - (2.92" - 0.18" - 0.18")^2] - (\pi/4)(0.5625")^2$$
$$= 17.61 \text{ in.}^2$$

Appendix F (Non-mandatory) Design for Creep and Critical Deflection Applications

F.1 Creep

F.1.1 Reference modulus of elasticity design values, E, in this Specification are intended for the calculation of immediate deformation under load. Under sustained loading, wood members exhibit additional time dependent deformation (creep) which usually develops at a slow but persistent rate over long periods of time. Creep rates are greater for members drying under load or exposed to varying temperature and relative humidity conditions than for members in a stable environment and at constant moisture content.

F.1.2 In certain bending applications, it may be necessary to limit deflection under long-term loading to specified levels. This can be done by applying an increase factor to the deflection due to long-term load. Total deflection is thus calculated as the immediate deflection due to the long-term component of the design load times the appropriate increase factor, plus the deflection due to the short-term or normal component of the design load.

F.2 Variation in Modulus of Elasticity

F.2.1 The reference modulus of elasticity design values, E, listed in Tables 4A, 4B, 4C, 4D, 4E, 4F, 5A, 5B, 5C, and 5D (published in the Supplement to this Specification) are average values and individual pieces having values both higher and lower than the averages will occur in all grades. The use of average modulus of elasticity values is customary practice for the design of normal wood structural members and assemblies. Field experience and tests have demonstrated that average values provide an adequate measure of the immediate deflection or deformation of these wood elements.

F.2.2 In certain applications where deflection may be critical, such as may occur in closely engineered, innovative structural components or systems, use of a reduced modulus of elasticity value may be deemed appropriate by the designer. The coefficient of variation in Table F1 shall be permitted to be used as a basis for modifying reference modulus of elasticity values listed in Tables 4A, 4B, 4C, 4D, 4E, 4F, 5A, 5B, 5C, and 5D to meet particular end use conditions.

F.2.3 Reducing reference average modulus of elasticity design values in this Specification by the product

of the average value and 1.0 and 1.65 times the applicable coefficients of variation in Table F1 gives estimates of the level of modulus of elasticity exceeded by 84% and 95%, respectively, of the individual pieces, as specified in the following formulas:

$$E_{0.16} = E(1 - 1.0 \, COV_E) \quad\quad (F-1)$$

$$E_{0.05} = E(1 - 1.645 \, COV_E) \quad\quad (F-2)$$

Table F1 Coefficients of Variation in Modulus of Elasticity (COV$_E$) for Lumber and Structural Glued Laminated Timber

	COV$_E$
Visually graded sawn lumber (Tables 4A, 4B, 4D, 4E, and 4F)	0.25
Machine Evaluated Lumber (MEL) (Table 4C)	0.15
Machine Stress Rated (MSR) lumber (Table 4C)	0.11
Structural Glued laminated timber (Tables 5A, 5B, 5C, and 5D)	0.10[1]

1. The COV$_E$ for structural glued laminated timber decreases as the number of laminations increases, and increases as the number of laminations decreases. COV$_E$ = 0.10 is an approximate value for six or more laminations.

F.3 Shear Deflection

F.3.1 Reference modulus of elasticity design values, E, listed in Tables 4A, 4B, 4C, 4D, 4E, 4F, 5A, 5B, 5C, and 5D are apparent modulus of elasticity values and include a shear deflection component. For sawn lumber, the ratio of shear-free E to reference E is 1.03. For structural glued laminated timber, the ratio of shear-free E to reference E is 1.05.

F.3.2 In certain applications use of an adjusted modulus of elasticity to more accurately account for the shear component of the total deflection may be deemed appropriate by the designer. Standard methods for adjusting modulus of elasticity to other load and span-depth conditions are available (see Reference 54). When reference modulus of elasticity values have not been adjusted to include the effects of shear deformation, such as for pre-fabricated wood I-joists, consideration for the shear component of the total deflection is required.

F.3.3 The shear component of the total deflection of a beam is a function of beam geometry, modulus of elasticity, shear modulus, applied load and support conditions. The ratio of shear-free E to apparent E is 1.03 for the condition of a simply supported rectangular beam with uniform load, a span to depth ratio of 21:1, and elastic modulus to shear modulus ratio of 16:1. The ratio of shear-free E to apparent E is 1.05 for a similar beam with a span to depth ratio of 17:1. See Reference 53 for information concerning calculation of beam deflection for other span-depth and load conditions.

Appendix G (Non-mandatory) Effective Column Length

G.1

The effective column length of a compression member is the distance between two points along its length at which the member is assumed to buckle in the shape of a sine wave.

G.2

The effective column length is dependent on the values of end fixity and lateral translation (deflection) associated with the ends of columns and points of lateral support between the ends of column. It is recommended that the effective length of columns be determined in accordance with good engineering practice. Lower values of effective length will be associated with more end fixity and less lateral translation while higher values will be associated with less end fixity and more lateral translation.

G.3

In lieu of calculating the effective column length from available engineering experience and methodology, the buckling length coefficients, K_e, given in Table G1 shall be permitted to be multiplied by the actual column length, ℓ, or by the length of column between lateral supports to calculate the effective column length, ℓ_e.

G.4

Where the bending stiffness of the frame itself provides support against buckling, the buckling length coefficient, K_e, for an unbraced length of column, ℓ, is dependent upon the amount of bending stiffness provided by the other in-plane members entering the connection at each end of the unbraced segment. If the combined stiffness from these members is sufficiently small relative to that of the unbraced column segments, K_e could exceed the values given in Table G1.

Table G1 Buckling Length Coefficients, K_e

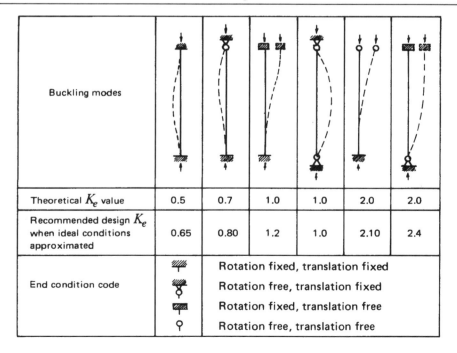

Buckling modes						
Theoretical K_e value	0.5	0.7	1.0	1.0	2.0	2.0
Recommended design K_e when ideal conditions approximated	0.65	0.80	1.2	1.0	2.10	2.4

End condition code		Rotation fixed, translation fixed
		Rotation free, translation fixed
		Rotation fixed, translation free
		Rotation free, translation free

Appendix H (Non-mandatory) Lateral Stability of Columns

H.1

Solid wood columns can be classified into three length classes, characterized by mode of failure at ultimate load. For short, rectangular columns with a small ratio of length to least cross-sectional dimension, ℓ_e/d, failure is by crushing. When there is an intermediate ℓ_e/d ratio, failure is generally a combination of crushing and buckling. At large ℓ_e/d ratios, long wood columns behave essentially as Euler columns and fail by lateral deflection or buckling. Design of these three length classes are represented by the single column Equation H-1.

H.2

For solid columns of rectangular cross section where the slenderness ratio, ℓ_e/d, does not exceed 50, adjusted compression design values parallel to grain are obtained by the equation:

$$F_c' = F_c^* \left[\frac{1+\left(F_{cE}/F_c^*\right)}{2c} - \sqrt{\left[\frac{1+\left(F_{cE}/F_c^*\right)}{2c}\right]^2 - \frac{F_{cE}/F_c^*}{c}} \right] \quad \text{(H-1)}$$

where:

$$F_{cE} = \frac{0.822\,E_{min}'}{\left(\ell_e/d\right)^2} \quad \text{(H-2)}$$

F_c^* = reference compression design value parallel to grain multiplied by all applicable adjustment factors except C_P (see 2.3)

c = 0.8 for sawn lumber

c = 0.85 for round timber poles and piles

c = 0.9 for structural glued laminated timber or structural composite lumber

Equation H-2 is derived from the standard Euler equation, with radius of gyration, r, converted to the more convenient least cross-sectional dimension, d, of a rectangular column.

H.3

The equation for adjusted compression design value, F_c', in this Specification is for columns having rectangular cross sections. It may be used for other column shapes by substituting $r\sqrt{12}$ for d in the equations, where r is the applicable radius of gyration of the column cross section.

H.4

The 0.822 factor in Equation H-2 represents the Euler buckling coefficient for rectangular columns calculated as $\pi^2/12$. Modulus of elasticity for beam and column stability, E_{min}, in Equation H-2 represents an approximate 5% lower exclusion value on pure bending modulus of elasticity, plus a 1.66 factor of safety (see Appendix D.4).

H.5

Adjusted design values based on Equations H-1 and H-2 are customarily used for most sawn lumber column designs. Where unusual hazard exists, a larger reduction factor may be appropriate. Alternatively, in less critical end use, the designer may elect to use a smaller factor of safety.

H.6

For products with less E variability than visually graded sawn lumber, higher critical buckling design values may be calculated. For a product having a lower coefficient of variation (COV_E), use of Equation H-2 will provide a 1.66 factor of safety at the 5% lower exclusion value.

Appendix I (Non-mandatory) Yield Limit Equations for Connections

I.1 Yield Modes

The yield limit equations specified in 11.3.1 for dowel-type fasteners such as bolts, lag screws, wood screws, nails, and spikes represent four primary connection yield modes (see Figure I1). Modes I_m and I_s represent bearing-dominated yield of the wood fibers in contact with the fastener in either the main or side member(s), respectively. Mode II represents pivoting of the fastener at the shear plane of a single shear connection with localized crushing of wood fibers near the faces of the wood member(s). Modes III_m and III_s represent fastener yield in bending at one plastic hinge point per shear plane, and bearing-dominated yield of wood fibers in contact with the fastener in either the main or side member(s), respectively. Mode IV represents fastener yield in bending at two plastic hinge points per shear plane, with limited localized crushing of wood fibers near the shear plane(s).

I.2 Dowel Bearing Strength for Steel Members

Dowel bearing strength, F_e, for steel members shall be based on accepted steel design practices (see References 39, 40 and 41). Design values in Tables 11B, 11D, 11G, 11I, 11J, 11M, and 11N are for 1/4" ASTM A 36 steel plate or 3 gage and thinner ASTM A 653, Grade 33 steel plate with dowel bearing strength proportional to ultimate tensile strength. Bearing strengths used to calculate connection yield load represent nominal bearing strengths of 2.4 F_u and 2.2 F_u, respectively (based on design provisions in References 39, 40, and 41 for bearing strength of steel members at connections). To allow proper application of the load duration factor for these connections, the bearing strengths have been divided by 1.6.

I.3 Dowel Bearing Strength for Wood Members

Dowel bearing strength, F_e, for wood members may be determined in accordance with ASTM D 5764.

I.4 Fastener Bending Yield Strength, F_{yb}

In the absence of published standards which specify fastener strength properties, the designer should contact fastener manufacturers to determine fastener bending yield strength for connection design. ASTM F 1575 provides a standard method for testing bending yield strength of nails.

Fastener bending yield strength (F_{yb}) shall be determined by the 5% diameter (0.05D) offset method of analyzing load-displacement curves developed from fastener bending tests. However, for short, large diameter fasteners for which direct bending tests are impractical, test data from tension tests such as those specified in ASTM F 606 shall be evaluated to estimate F_{yb}.

Research indicates that F_{yb} for bolts is approximately equivalent to the average of bolt tensile yield strength and bolt tensile ultimate strength, $F_{yb} = F_y/2 + F_u/2$. Based on this approximation, 48,000 psi $\leq F_{yb} \leq$ 140,000 psi for various grades of SAE J429 bolts. Thus, the aforementioned research indicates that $F_{yb} = 45,000$ psi is reasonable for many commonly available bolts. Tests of limited samples of lag screws indicate that $F_{yb} = 45,000$ psi is also reasonable for many commonly available lag screws with D $\geq 3/8"$.

Tests of a limited sample of box nails and common wire nails from twelve U.S. nail manufacturers indicate that F_{yb} increases with decreasing nail diameter, and may exceed 100,000 psi for very small diameter nails. These tests indicate that the F_{yb} values used in Tables 11N through 11R are reasonable for many commonly available box nails and small diameter common wire nails (D < 0.2"). Design values for large diameter common wire nails (D > 0.2") are based on extrapolated estimates of F_{yb} from the aforementioned limited study. For hardened-steel nails, F_{yb} is assumed to be approximately 30% higher than for the same diameter common wire nails. Design values in Tables 11J through 11M for wood screws and small diameter lag screws (D < 3/8") are based on estimates of F_{yb} for common wire nails of the same diameter. Table I1 provides values of F_{yb} based on fastener type and diameter.

Figure I1 (Non-mandatory) Connection Yield Modes

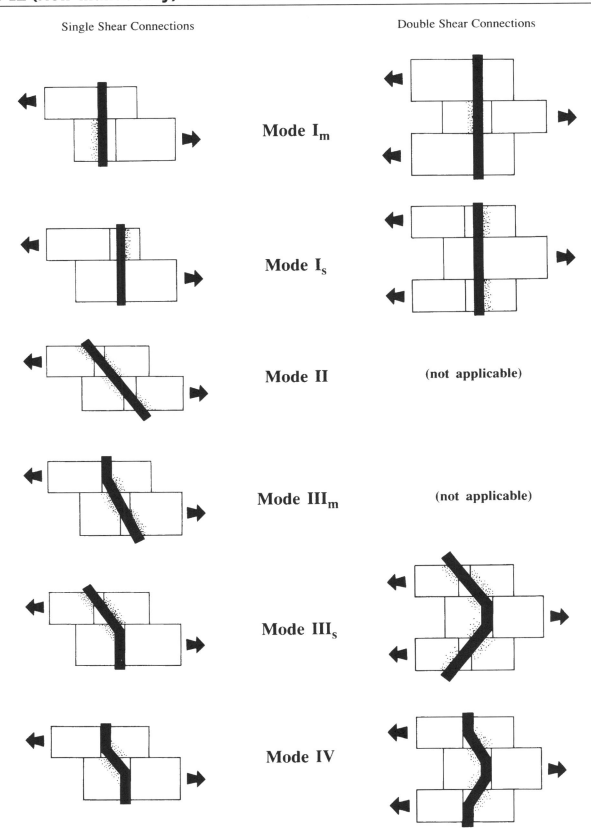

I.5 Threaded Fasteners

The reduced moment resistance in the threaded portion of dowel-type fasteners can be accounted for by use of root diameter, D_r, in calculation of reference lateral design values. Use of diameter, D, is permitted when the threaded portion of the fastener is sufficiently far away from the connection shear plane(s). For example, diameter, D, may be used when the length of thread bearing in the main member of a two member connection does not exceed 1/4 of the total bearing length in the main member (member holding the threads). For a connection with three or more members, diameter, D, may be used when the length of thread bearing in the outermost member does not exceed 1/4 of the total bearing length in the outermost member (member holding the threads). Use of diameter, D, is permitted when full body diameter bolts defined in 11.1.2 are used since thread bearing lengths are typically small in the member holding the threads. For thread lengths greater than 1/4 of the total bearing length in the member holding the threads, the effect of reduced moment resistance

of bolts defined in 11.1.2 is small when evaluated with a more detailed analysis.

Reference lateral design values for reduced body diameter lag screw and rolled thread wood screw connections are based on root diameter, D_r to account for the reduced diameter of these fasteners. These values may also be applicable for full-body diameter lag screws and cut thread wood screws since the length of threads for these fasteners is generally not known and/or the thread bearing length based on typical dimensions exceeds 1/4 the total bearing length in the member holding the threads. For bolted connections, reference lateral design values are based on diameter, D.

One alternate method of accounting for the moment and bearing resistance of the threaded portion of the fastener and moment acting along the length of the fastener is provided in AF&PA's *Technical Report 12 - General Dowel Equations for Calculating Lateral Connection Values* (see Reference 51). A general set of equations permits use of different fastener diameters for bearing resistance and moment resistance in each member.

Table I1   Fastener Bending Yield Strengths, F_{yb}

Fastener Type	F_{yb} (psi)
Bolt, lag screw (with D ≥ 3/8"), drift pin (SAE J429 Grade 1 - F_y = 36,000 psi and F_u = 60,000 psi)	45,000
Common, box, or sinker nail, spike, lag screw, wood screw (low to medium carbon steel)	
0.099" ≤ D ≤ 0.142"	100,000
0.142" < D ≤ 0.177"	90,000
0.177" < D ≤ 0.236"	80,000
0.236" < D ≤ 0.273"	70,000
0.273" < D ≤ 0.344"	60,000
0.344" < D ≤ 0.375"	45,000
Hardened steel nail (medium carbon steel)	
0.120" ≤ D ≤ 0.142"	130,000
0.142" < D ≤ 0.192"	115,000
0.192" < D ≤ 0.207"	100,000

Appendix J (Non-mandatory) Solution of Hankinson Formula

J.1

When members are loaded in bearing at an angle to grain between 0° and 90°, or when split ring or shear plate connectors, bolts, or lag screws are loaded at an angle to grain between 0° and 90°, design values at an angle to grain shall be determined using the Hankinson formula.

J.2

The Hankinson formula is for the condition where the loaded surface is perpendicular to the direction of the applied load.

J.3

When the resultant force is not perpendicular to the surface under consideration, the angle θ is the angle between the direction of grain and the direction of the force component which is perpendicular to the surface.

J.4

The bearing surface for a split ring or shear plate connector, bolt or lag screw is assumed perpendicular to the applied lateral load.

J.5

The bearing strength of wood depends upon the direction of grain with respect to the direction of the applied load. Wood is stronger in compression parallel to grain than in compression perpendicular to grain. The variation in strength at various angles to grain between 0° and 90° shall be determined by the Hankinson formula as follows:

$$F_\theta' = \frac{F_c^* F_{c\perp}'}{F_c^* \sin^2 \theta + F_{c\perp}' \cos^2 \theta} \qquad (J\text{-}1)$$

where:

F_c^* = adjusted compression design value parallel to grain multiplied by all applicable adjustment factors except the column stability factor

$F_{c\perp}'$ = adjusted compression design value perpendicular to grain

F_θ' = adjusted bearing design value at an angle to grain

θ = angle between direction of load and direction of grain (longitudinal axis of member)

When determining dowel bearing design values at an angle to grain for bolt or lag screw connections, the Hankinson formula takes the following form:

$$F_{e\theta} = \frac{F_{e\parallel} F_{e\perp}}{F_{e\parallel} \sin^2 \theta + F_{e\perp} \cos^2 \theta} \qquad (J\text{-}2)$$

where:

$F_{e\parallel}$ = dowel bearing strength parallel to grain

$F_{e\perp}$ = dowel bearing strength perpendicular to grain

$F_{e\theta}$ = dowel bearing strength at an angle to grain

When determining adjusted design values for bolt or lag screw wood-to-metal connections or wood-to-wood connections with the main or side member(s) loaded parallel to grain, the following form of the Hankinson formula provides an alternate solution:

$$Z_\theta' = \frac{Z_\parallel' Z_\perp'}{Z_\parallel' \sin^2 \theta + Z_\perp' \cos^2 \theta} \qquad (J\text{-}3)$$

For wood-to-wood connections with side member(s) loaded parallel to grain,

$Z_\parallel'$ = adjusted lateral design value for a single bolt or lag screw connection with the main and side wood members loaded parallel to grain, $Z_\parallel$

$Z_\perp'$ = adjusted lateral design value for a single bolt or lag screw connection with the side member(s) loaded parallel to grain and main member loaded perpendicular to grain, $Z_{m\perp}$

For wood-to-wood connections with the main member loaded parallel to grain,

$Z_\parallel'$ = adjusted lateral design value for a single bolt or lag screw connection with the main and side wood members loaded parallel to grain, $Z_\parallel$

$Z_\perp'$ = adjusted lateral design value for a single bolt or lag screw connection with the main member loaded parallel to grain and side member(s) loaded perpendicular to grain, $Z_{s\perp}$

For wood-to-metal connections,

$Z_\parallel'$ = adjusted lateral design value for a single bolt or lag screw connection with the wood member loaded parallel to grain, $Z_\parallel$

$Z_\perp'$ = adjusted lateral design value for a single bolt or lag screw connection with the wood member loaded perpendicular to grain, $Z_\perp$

When determining adjusted design values for split ring or shear plate connectors or timber rivets, the Hankinson formula takes the following form:

$$N' = \frac{P'Q'}{P'\sin^2\theta + Q'\cos^2\theta} \qquad (J\text{-}4)$$

where:

P' = adjusted lateral design value parallel to grain for a single split ring connector unit or shear plate connector unit

Q' = adjusted lateral design value perpendicular to grain for a single split ring connector unit or shear plate connector unit

N' = adjusted lateral design value at an angle to grain for a single split ring connector unit or shear plate connector unit

The nomographs presented in Figure J1 provide a graphical solution of the Hankinson formula.

Figure J1 Solution of Hankinson Formula

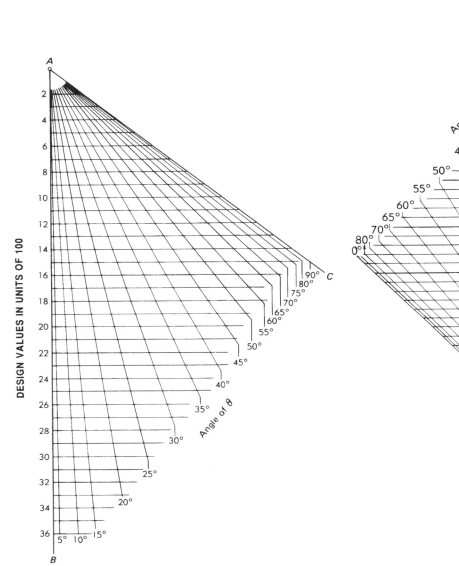

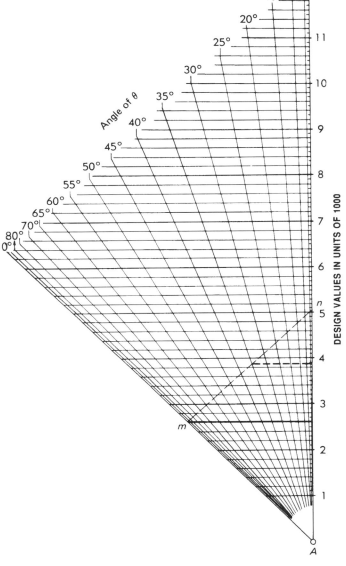

Figure J2   Connection Loaded at an Angle to Grain

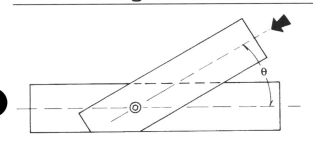

Sample Solution for Split Ring or Shear Plate Connection:

Assume that P' = 5,030 lbs., Q' = 2,620 lbs., and θ = 35° in Figure J2. On line A-B in Figure J1, locate 5,030 lbs. at point n. On the same line A-B, locate 2,620 lbs. and project to point m on line A-C. Where line m-n intersects the radial line for 35°, project to line A-B and read the ASD adjusted design value, N' = 3,870 lbs.

Appendix K (Non-mandatory) Typical Dimensions for Split Ring and Shear Plate Connectors

SPLIT RINGS[1]	2-1/2"	4"
Split Ring		
Inside diameter at center when closed	2.500"	4.000"
Thickness of metal at center	0.163"	0.193"
Depth of metal (width of ring)	0.750"	1.000"
Groove		
Inside diameter	2.56"	4.08"
Width	0.18"	0.21"
Depth	0.375"	0.50"
Bolt hole diameter in timber members	9/16"	13/16"
Washers, standard		
Round, cast or malleable iron, diameter	2-1/8"	3"
Round, wrought iron (minimum)		
Diameter	1-3/8"	2"
Thickness	3/32"	5/32"
Square plate		
Length of side	2"	3"
Thickness	1/8"	3/16"
Projected area: portion of one split ring within member	1.10 in.^{2}	2.24 in.^{2}

1. Courtesy of Cleveland Steel Specialty Co.

SHEAR PLATES	2-5/8"	2-5/8"	4"	4"
Shear plate[1]	Pressed	Malleable	Malleable	Malleable
Material	steel	cast iron	cast iron	cast iron
Plate diameter	2.62"	2.62"	4.02"	4.02"
Bolt hole diameter	0.81"	0.81"	0.81"	0.93"
Plate thickness	0.172"	0.172"	0.20"	0.20"
Plate depth	0.42"	0.42"	0.62"	0.62"
Bolt hole diameter in timber members and metal side plates[2]	13/16"	13/16"	13/16"	15/16"
Washers, standard				
Round, cast or malleable iron, diameter	3"	3"	3"	3-1/2"
Round, wrought iron (minimum)				
Diameter	2"	2"	2"	2-1/4"
Thickness	5/32"	5/32"	5/32"	11/64"
Square plate				
Length of side	3"	3"	3"	3"
Thickness	1/4"	1/4"	1/4"	1/4"
Projected area: portion of one shear plate within member	1.18 in.^{2}	1.00 in.^{2}	2.58 in.^{2}	2.58 in.^{2}

1. ASTM D 5933.

2. Steel straps or shapes used as metal side plates shall be designed in accordance with accepted metal practices (see 10.2.3).

Appendix L (Non-mandatory) Typical Dimensions for Dowel-Type Fasteners[1]

Table L1   Standard Hex Bolts

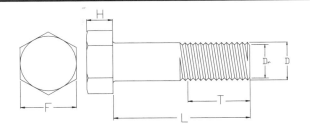

D = diameter
D_r = root diameter
T = thread length
L = bolt length
F = width of head across flats
H = height of head

		Diameter, D							
		1/4"	**5/16"**	**3/8"**	**1/2"**	**5/8"**	**3/4"**	**7/8"**	**1"**
D_r		0.189"	0.245"	0.298"	0.406"	0.514"	0.627"	0.739"	0.847"
F		7/16"	1/2"	9/16"	3/4"	15/16"	1-1/8"	1-5/16"	1-1/2"
H		11/64"	7/32"	1/4"	11/32"	27/64"	1/2"	37/64"	43/64"
T	L ≤ 6 in.	3/4"	7/8"	1"	1-1/4"	1-1/2"	1-3/4"	2"	2-1/4"
	L > 6 in.	1"	1-1/8"	1-1/4"	1-1/2"	1-3/4"	2"	2-1/4"	2-1/2"

1. Tolerances specified in ANSI B 18.2.1. Full body diameter bolt is shown. Root diameter based on UNC (coarse) thread series (see ANSI B1.1).

APPENDIX

A

HEX BOLTS

Table L2   Standard Hex Lag Screws[1]

D = diameter
D_r = root diameter
S = unthreaded shank length
T = minimum thread length[2]

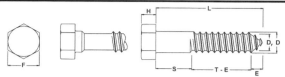

Reduced Body Diameter   Full Body Diameter

E = length of tapered tip
N = number of threads/inch
F = width of head across flats
H = height of head

Length, L		Diameter, D										
		1/4"	5/16"	3/8"	7/16"	1/2"	5/8"	3/4"	7/8"	1"	1-1/8"	1-1/4"
	D_r	0.173"	0.227"	0.265"	0.328"	0.371"	0.471"	0.579"	0.683"	0.780"	0.887"	1.012"
	E	5/32"	3/16"	7/32"	9/32"	5/16"	13/32"	1/2"	19/32"	11/16"	25/32"	7/8"
	H	11/64"	7/32"	1/4"	19/64"	11/32"	27/64"	1/2"	37/64"	43/64"	3/4"	27/32"
	F	7/16"	1/2"	9/16"	5/8"	3/4"	15/16"	1-1/8"	1-5/16"	1-1/2"	1-11/16"	1-7/8"
	N	10	9	7	7	6	5	4-1/2	4	3-1/2	3-1/4	3-1/4
1"	S	1/4"	1/4"	1/4"	1/4"	1/4"						
	T	3/4"	3/4"	3/4"	3/4"	3/4"						
	T-E	19/32"	9/16"	17/32"	15/32"	7/16"						
1-1/2"	S	1/4"	1/4"	1/4"	1/4"	1/4"						
	T	1-1/4"	1-1/4"	1-1/4"	1-1/4"	1-1/4"						
	T-E	1-3/32"	1-1/16"	1-1/32"	31/32"	15/16"						
2"	S	1/2"	1/2"	1/2"	1/2"	1/2"	1/2"					
	T	1-1/2"	1-1/2"	1-1/2"	1-1/2"	1-1/2"	1-1/2"					
	T-E	1-11/32"	1-5/16"	1-9/32"	1-7/32"	1-3/16"	1-3/32"					
2-1/2"	S	3/4"	3/4"	3/4"	3/4"	3/4"	3/4"					
	T	1-3/4"	1-3/4"	1-3/4"	1-3/4"	1-3/4"	1-3/4"					
	T-E	1-19/32"	1-9/16"	1-17/32"	1-15/32"	1-7/16"	1-11/32"					
3	S	1"	1"	1"	1"	1"	1"	1"	1"	1"		
	T	2"	2"	2"	2"	2"	2"	2"	2"	2"		
	T-E	1-27/32"	1-13/16"	1-25/32"	1-23/32"	1-11/16"	1-19/32"	1-1/2"	1-13/32"	1-5/16"		
4"	S	1-1/2"	1-1/2"	1-1/2"	1-1/2"	1-1/2"	1-1/2"	1-1/2"	1-1/2"	1-1/2"	1-1/2"	1-1/2"
	T	2-1/2"	2-1/2"	2-1/2"	2-1/2"	2-1/2"	2-1/2"	2-1/2"	2-1/2"	2-1/2"	2-1/2"	2-1/2"
	T-E	2-11/32"	2-5/16"	2-9/32"	2-7/32"	2-3/16"	2-3/32"	2"	1-29/32"	1-13/16"	1-23/32"	1-5/8"
5"	S	2"	2"	2"	2"	2"	2"	2"	2"	2"	2"	2"
	T	3"	3"	3"	3"	3"	3"	3"	3"	3"	3"	3"
	T-E	2-27/32"	2-13/16"	2-25/32"	2-23/32"	2-11/16"	2-19/32"	2-1/2"	2-13/32"	2-5/16"	2-7/32"	2-1/8"
6"	S	2-1/2"	2-1/2"	2-1/2"	2-1/2"	2-1/2"	2-1/2"	2-1/2"	2-1/2"	2-1/2"	2-1/2"	2-1/2"
	T	3-1/2"	3-1/2"	3-1/2"	3-1/2"	3-1/2"	3-1/2"	3-1/2"	3-1/2"	3-1/2"	3-1/2"	3-1/2"
	T-E	3-11/32"	3-5/16"	3-9/32"	3-7/32"	3-3/16"	3-3/32"	3"	2-29/32"	2-13/16"	2-23/32"	2-5/8"
7"	S	3"	3"	3"	3"	3"	3"	3"	3"	3"	3"	3"
	T	4"	4"	4"	4"	4"	4"	4"	4"	4"	4"	4"
	T-E	3-27/32"	3-13/16"	3-25/32"	3-23/32"	3-11/16"	3-19/32"	3-1/2"	3-13/32"	3-5/16"	3-7/32"	3-1/8"
8"	S	3-1/2"	3-1/2"	3-1/2"	3-1/2"	3-1/2"	3-1/2"	3-1/2"	3-1/2"	3-1/2"	3-1/2"	3-1/2"
	T	4-1/2"	4-1/2"	4-1/2"	4-1/2"	4-1/2"	4-1/2"	4-1/2"	4-1/2"	4-1/2"	4-1/2"	4-1/2"
	T-E	4-11/32"	4-5/16"	4-9/32"	4-7/32"	4-3/16"	4-3/32"	4"	3-29/32"	3-13/16"	3-23/32"	3-5/8"
9"	S	4"	4"	4"	4"	4"	4"	4"	4"	4"	4"	4"
	T	5"	5"	5"	5"	5"	5"	5"	5"	5"	5"	5"
	T-E	4-27/32"	4-13/16"	4-25/32"	4-23/32"	4-11/16"	4-19/32"	4-1/2"	4-13/32"	4-5/16"	4-7/32"	4-1/8"
10"	S	4-1/2"	4-1/2"	4-1/2"	4-1/2"	4-1/2"	4-1/2"	4-1/2"	4-1/2"	4-1/2"	4-1/2"	4-1/2"
	T	5-1/2"	5-1/2"	5-1/2"	5-1/2"	5-1/2"	5-1/2"	5-1/2"	5-1/2"	5-1/2"	5-1/2"	5-1/2"
	T-E	5-11/32"	5-5/16"	5-9/32"	5-7/32"	5-3/16"	5-3/32"	5"	4-29/32"	4-13/16"	4-23/32"	4-5/8"
11"	S	5"	5"	5"	5"	5"	5"	5"	5"	5"	5"	5"
	T	6"	6"	6"	6"	6"	6"	6"	6"	6"	6"	6"
	T-E	5-27/32"	5-13/16"	5-25/32"	5-23/32"	5-11/16"	5-19/32"	5-1/2"	5-13/32"	5-5/16"	5-7/32"	5-1/8"
12"	S	6"	6"	6"	6"	6"	6"	6"	6"	6"	6"	6"
	T	6"	6"	6"	6"	6"	6"	6"	6"	6"	6"	6"
	T-E	5-27/32"	5-13/16"	5-25/32"	5-23/32"	5-11/16"	5-19/32"	5-1/2"	5-13/32"	5-5/16"	5-7/32"	5-1/8"

1. Tolerances specified in ANSI B18.2.1. Full body diameter and reduced body diameter lag screws are shown. For reduced body diameter lag screws, the unthreaded shank diameter may be reduced to approximately the root diameter, D_r.

2. Minimum thread length (T) for lag screw lengths (L) is 6" or 1/2 the lag screw length plus 0.5", whichever is less. Thread lengths may exceed these minimums up to the full lag screw length (L).

HEX LAG SCREWS

Table L3   Standard Wood Screws[1]

 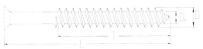

D = diameter
D$_r$ = root diameter
L = screw length
T = thread length

Cut Thread[2]   Rolled Thread[3]

	Wood Screw Number										
	6	7	8	9	10	12	14	16	18	20	24
D	0.138"	0.151"	0.164"	0.177"	0.19"	0.216"	0.242"	0.268"	0.294"	0.32"	0.372"
D$_r$[4]	0.113"	0.122"	0.131"	0.142"	0.152"	0.171"	0.196"	0.209"	0.232"	0.255"	0.298"

1. Tolerances specified in ANSI B18.6.1
2. Thread length on cut thread wood screws is approximately 2/3 of the screw length.
3. Single lead thread shown. Thread length is at least four times the screw diameter or 2/3 of the screw length, whichever is greater. Screws which are too short to accommodate the minimum thread length, have threads extending as close to the underside of the head as practicable.
4. Taken as the average of the specified maximum and minimum limits for body diameter of rolled thread wood screws.

Table L4   Standard Common, Box, and Sinker Nails[1]

D = diameter
L = length
H = head diameter

Common or Box   Sinker

Type		Pennyweight										
		6d	7d	8d	10d	12d	16d	20d	30d	40d	50d	60d
Common	L	2"	2-1/4"	2-1/2"	3"	3-1/4"	3-1/2"	4"	4-1/2"	5"	5-1/2"	6"
	D	0.113"	0.113"	0.131"	0.148"	0.148"	0.162"	0.192"	0.207"	0.225"	0.244"	0.263"
	H	0.266"	0.266"	0.281"	0.312"	0.312"	0.344"	0.406"	0.438"	0.469"	0.5"	0.531"
Box	L	2"	2-1/4"	2-1/2"	3"	3-1/4"	3-1/2"	4"	4-1/2"	5"		
	D	0.099"	0.099"	0.113"	0.128"	0.128"	0.135"	0.148"	0.148"	0.162"		
	H	0.266"	0.266"	0.297"	0.312"	0.312"	0.344"	0.375"	0.375"	0.406"		
Sinker	L	1-7/8"	2-1/8"	2-3/8"	2-7/8"	3-1/8"	3-1/4"	3-3/4"	4-1/4"	4-3/4"		5-3/4"
	D	0.092"	0.099"	0.113"	0.12"	0.135"	0.148"	0.177"	0.192"	0.207"		0.244"
	H	0.234"	0.250"	0.266"	0.281"	0.312"	0.344"	0.375"	0.406"	0.438"		0.5"

1. Tolerances specified in ASTM F 1667. Typical shape of common, box, and sinker nails shown. See ASTM F 1667 for other nail types.

A

APPENDIX

WOOD SCREWS AND NAILS

Appendix M (Non-mandatory) Manufacturing Tolerances for Rivets and Steel Side Plates for Timber Rivet Connections

Rivet dimensions are taken from ASTM F 1667.

Rivet Dimensions

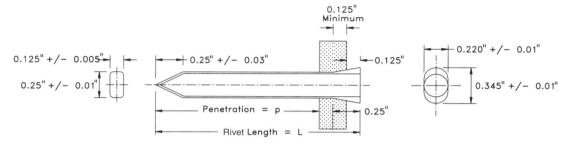

Steel Side Plate Dimensions

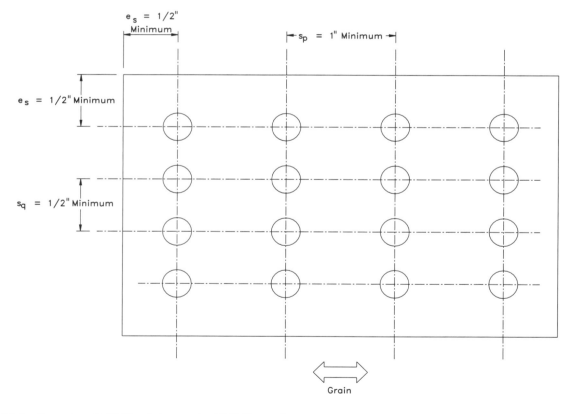

Notes:

1. Hole diameter: 17/64" minimum to 18/64" maximum.
2. Tolerences in location of holes: 1/8" maximum in any direction.
3. All dimensions are prior to galvanizing in inches.
4. s_p and s_q are defined in 13.3.
5. e_s is the end and edge distance as defined by the steel.
6. Orient wide face of rivets parallel to grain, regardless of plate orientation.

Appendix N (Mandatory) Load and Resistance Factor Design (LRFD)

N.1 General

N.1.1 Application

LRFD designs shall be made in accordance with Appendix N and all applicable provisions of this Specification. Applicable loads and load combinations, and adjustment of design values unique to LRFD are specified herein.

N.1.2 Loads and Load Combinations

Nominal loads and load combinations shall be those required by the applicable building code. In the absence of a governing building code, the nominal loads and associated load combinations shall be those specified in ASCE 7.

N.2 Design Values

N.2.1 Design Values

Adjusted LRFD design values for members and connections shall be determined in accordance with ASTM Specification D 5457 and design provisions in this Specification or in accordance with N.2.2 and N.2.3. Where LRFD design values are determined by the reliability normalization factor method in ASTM D 5457, the format conversion factor shall not apply (see N.3.1).

N.2.2 Member Design Values

Reference member design values in this Specification shall be adjusted in accordance with 4.3, 5.3, 6.3,

7.3, 8.3, and 9.3 for sawn lumber, structural glued laminated timber, poles and piles, prefabricated wood I-joists, structural composite lumber, and panel products, respectively, to determine the adjusted LRFD design value.

N.2.3 Connection Design Values

Reference connection design values in this Specification shall be adjusted in accordance with Table 10.3 to determine the adjusted LRFD design value.

N.3 Adjustment of Reference Design Values

N.3.1 Format Conversion Factor, K_F (LRFD Only)

Reference design values shall be multiplied by the format conversion factor, K_F, as specified in Table N1. Format conversion factors in Table N1 adjust reference

ASD design values (based on normal duration) to the LRFD reference resistances (see Reference 55). Format conversion factors shall not apply where LRFD reference resistances are determined in accordance with the reliability normalization factor method in ASTM D 5457.

Table N1 Format Conversion Factor, K_F (LRFD Only)

Application	Property	K_F
Member	F_b, F_t, F_v, F_c, F_{rt}, F_s	$2.16/\phi$
	$F_{c\perp}$	$1.875/\phi$
	E_{min}	$1.5/\phi$
Connections	(all connections in the *NDS*)	$2.16/\phi$

N.3.2 Resistance Factor, ϕ (LRFD Only)

Reference design values shall be multiplied by the resistance factor, ϕ, as specified in Table N2 (see Reference 55).

Table N2 Resistance Factor, ϕ (LRFD Only)

Application	Property	Symbol	Value
Member	F_b	ϕ_b	0.85
	F_t	ϕ_t	0.80
	F_v, F_{rt}, F_s	ϕ_v	0.75
	F_c, $F_{c\perp}$	ϕ_c	0.90
	E_{min}	ϕ_s	0.85
Connections	(all)	ϕ_z	0.65

N.3.3 Time Effect Factor, λ (LRFD Only)

Reference design values shall be multiplied by the time effect factor, λ, as specified in Table N3.

Table N3 Time Effect Factor, λ (LRFD Only)

Load Combination[2]	λ
$1.4(D+F)$	0.6
$1.2(D+F) + 1.6(H) + 0.5(L_r \text{ or } S \text{ or } R)$	0.6
$1.2(D+F) + 1.6(L+H) + 0.5(L_r \text{ or } S \text{ or } R)$	0.7 when L is from storage
	0.8 when L is from occupancy
	1.25 when L is from impact[1]
$1.2D + 1.6(L_r \text{ or } S \text{ or } R) + (L \text{ or } 0.8W)$	0.8
$1.2D + 1.6W + L + 0.5(L_r \text{ or } S \text{ or } R)$	1.0
$1.2D + 1.0E + L + 0.2S$	1.0
$0.9D + 1.6W + 1.6H$	1.0
$0.9D + 1.0E + 1.6H$	1.0

1. Time effect factors, λ, greater than 1.0 shall not apply to connections or to structural members pressure-treated with water-borne preservatives (see Reference 30) or fire retardant chemicals.
2. Load combinations and load factors consistent with ASCE 7-02 are listed for ease of reference. Nominal loads shall be in accordance with N.1.2.

REFERENCES

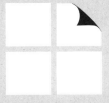

1. ACI 318-02 Building Code Requirements for Structural Concrete, American Concrete Institute, Farmington Hills, MI, 2002.

2. ACI 530-99/ASCE 5-99/TMS 402-99 Building Code Requirements for Masonry Structures, American Concrete Institute, Farmington Hills, MI, 1999.

3. AISI 1035 Standard Steels, American Iron and Steel Institute, Washington, DC, 1985.

4. ANSI/AITC Standard A190.1-2002, Structural Glued Laminated Timber, American Institute of Timber Construction, Vancouver, WA, 1992.

5. ANSI/ASCE Standard 7-02, Minimum Design Loads for Buildings and Other Structures, American Society of Civil Engineers, Reston, VA, 2003.

6. ANSI/ASME Standard B1.1-1989, Unified Inch Screw Threads UN and UNR Thread Form, American Society of Mechanical Engineers, New York, NY, 1989.

7. ANSI/ASME Standard B18.2.1-1996, Square and Hex Bolts and Screws (Inch Series), American Society of Mechanical Engineers, New York, NY, 1997.

8. ANSI/ASME Standard B18.6.1-1981 (Reaffirmed 1997), Wood Screws (Inch Series), American Society of Mechanical Engineers, New York, NY, 1982.

9. ANSI/TPI 1-2002 National Design Standard for Metal Plate Connected Wood Trusses, Truss Plate Institute, 2002.

10. ASTM Standard A 36-04, Specification for Standard Structural Steel, ASTM, West Conshohocken, PA, 2004.

11. ASTM Standard A 47-99, Specification for Ferritic Malleable Iron Castings, ASTM, West Conshohocken, PA, 1999.

12. ASTM A 153-03, Specification for Zinc Coating (Hot-Dip) on Iron and Steel Hardware, ASTM, West Conshohocken, PA, 2003.

13. ASTM A 370-03a, Standard Test Methods and Definitions for Mechanical Testing of Steel Products, ASTM, West Conshohocken, PA, 2003.

14. ASTM Standard A 653-03, Specification for Steel Sheet, Zinc-Coated (Galvanized) or Zinc-Iron Alloy-Coated (Galvannealed) by the Hot-Dip Process, 2003.

15. ASTM Standard D 25-91, Round Timber Piles, ASTM, West Conshohocken, PA, 1991.

16. ASTM Standard D 245-00$^{\varepsilon 1}$ (2002), Establishing Structural Grades and Related Allowable Properties for Visually Graded Lumber, ASTM, West Conshohocken, PA, 2002.

17. ASTM Standard D 1760-01, Pressure Treatment of Timber Products, ASTM, West Conshohocken, PA, 2001.

18. ASTM Standard D 1990-00$^{\varepsilon 1}$ (2002), Establishing Allowable Properties for Visually Graded Dimension Lumber from In-Grade Tests of Full-Size Specimens, ASTM, West Conshohocken, PA, 2002.

19. ASTM Standard D 2555-98$^{\varepsilon 1}$, Establishing Clear Wood Strength Values, ASTM, West Conshohocken, PA, 1998.

20. ASTM Standard D 2899-95, Establishing Design Stresses for Round Timber Piles, ASTM, West Conshohocken, PA, 1995.

21. ASTM Standard D 3200-74(2000), Establishing Recommended Design Stresses for Round Timber Construction Poles, ASTM, West Conshohocken, PA, 2000.

22. ASTM Standard D 3737-03, Establishing Stresses for Structural Glued Laminated Timber (Glulam), ASTM, West Conshohocken, PA, 2003.

23. ASTM Standard D 5055-04, Establishing and Monitoring Structural Capacities of Prefabricated Wood I-Joists, ASTM, West Conshohocken, PA, 2004.

24. ASTM Standard D 5456-03, Evaluation of Structural Composite Lumber Products, ASTM, West Conshohocken, PA, 2003.

25. ASTM Standard D 5764-97a (2002), Test Method for Evaluating Dowel Bearing Strength of Wood and Wood-Base Products, ASTM, West Conshohocken, PA, 2002.

26. ASTM Standard D 5933-96 (2001), Standard Specification for 2-5/8 in. and 4 in. Diameter Metal Shear Plates for Use in Wood Construction, ASTM, West Conshohocken, PA, 2001.

27. ASTM Standard F 606-02 [ε1], Determining the Mechanical Properties of Externally and Internally Threaded Fasteners, Washers, and Rivets, ASTM, West Conshohocken, PA, 2002.

28. ASTM Standard F 1575-03, Standard Test Method for Determining Bending Yield Moment of Nails, ASTM, West Conshohocken, PA, 2003.

29. ASTM Standard F 1667-03, Standard for Driven Fasteners: Nails, Spikes, and Staples, ASTM, West Conshohocken, PA, 2003.

30. AWPA Book of Standards, American Wood Preservers' Association, Selma, AL, 2003.

31. American Softwood Lumber Standard, Voluntary Product Standard PS 20-99, National Institute of Standards and Technology, U.S. Department of Commerce, 1999.

32. Design/Construction Guide-Diaphragms and Shear Walls, Form L350, APA-The Engineered Wood Association, Tacoma, WA, 2001.

33. Engineered Wood Construction Guide, Form E30, APA-The Engineered Wood Association, Tacoma, WA, 2001.

34. Plywood Design Specification and Supplements, Form Y510, APA-The Engineered Wood Association, Tacoma, WA, 1997.

35. PS1-95, Construction and Industrial Plywood, United States Department of Commerce, National Institute of Standards and Technology, Gaithersburg, MD, 1995.

36. PS2-92, Performance Standard for Wood-Based Structural-Use Panels, United States Department of Commerce, National Institute of Standards and Technology, Gaithersburg, MD, 1992.

37. SAE J412, General Characteristics and Heat Treatment of Steels, Society of Automotive Engineers, Warrendale, PA, 1995.

38. SAE J429, Mechanical and Material Requirements for Externally Threaded Fasteners, Society of Automotive Engineers, Warrendale, PA, 1999.

39. Specification for Structural Joints Using ASTM A325 or A490 Bolts, American Institute of Steel Construction (AISC), Chicago, IL, 1985.

40. Specification for Structural Steel Buildings–Allowable Stress Design and Plastic Design, American Institute of Steel Construction (AISC), Chicago, IL, 1989.

41. Specification for the Design of Cold-Formed Steel Structural Members, American Iron and Steel Institute (AISI), Washington, DC, 1996.

42. Standard Grading Rules for Canadian Lumber, National Lumber Grades Authority (NLGA), New Westminster, BC, Canada, 2003.

43. Standard Grading Rules for Northeastern Lumber, Northeastern Lumber Manufacturers Association (NELMA), Cumberland Center, ME, 2003.

44. Standard Grading Rules for Northern and Eastern Lumber, Northern Softwood Lumber Bureau (NSLB), Cumberland Center, ME, 1993.

45. Standard Grading Rules for Southern Pine Lumber, Southern Pine Inspection Bureau (SPIB), Pensacola, FL, 2002.

46. Standard Grading Rules for West Coast Lumber, West Coast Lumber Inspection Bureau (WCLIB), Portland, OR, 2004.

47. Standard Specifications for Grades of California Redwood Lumber, Redwood Inspection Service (RIS), Novato, CA, 2000.

48. Standard Specifications for Highway Bridges, American Association of State Highway and Transportation Officials (AASHTO), Washington, DC, 1987.

49. Western Lumber Grading Rules, Western Wood Products Association (WWPA), Portland, OR, 2005.

50. Design Manual for TECO Timber Connectors Construction, TECO/Lumberlok, Colliers, WV, 1973.

51. Technical Report 12 General Dowel Equations for Calculating Lateral Connection Values, American Forest & Paper Association (AF&PA), Washington, DC, 1999.

52. Timber Construction Manual, American Institute of Timber Construction (AITC), John Wiley & Sons, 2004.

53. Wood Handbook: Wood as an Engineering Material, General Technical Report 113, Forest Products Laboratory, U.S. Department of Agriculture, 1999.

54. ASTM Standard D 2915-03, Standard Practice for Evaluating Allowable Properties for Grades of Structural Lumber, ASTM West Conshohocken, PA, 2003.

55. ASTM Standard D 5457-04, Standard Specification for Computing the Reference Resistance of Wood-Based Materials and Structural Connections for Load and Resistance Factor Design, ASTM, West Conshohocken, PA, 2004.